Frontiers in Applied Dynamical Systems:
Reviews and Tutorials

Volume 1

More information about this series at http://www.springer.com/series/13763

Frontiers in Applied Dynamical Systems: Reviews and Tutorials

The Frontiers in Applied Dynamical Systems (FIADS) covers emerging topics and significant developments in the field of applied dynamical systems. It is a collection of invited review articles by leading researchers in dynamical systems, their applications and related areas. Contributions in this series should be seen as a portal for a broad audience of researchers in dynamical systems at all levels and can serve as advanced teaching aids for graduate students. Each contribution provides an informal outline of a specific area, an interesting application, a recent technique, or a "how-to" for analytical methods and for computational algorithms, and a list of key references. All articles will be refereed.

Richard Bertram • Joël Tabak • Wondimu Teka
Theodore Vo • Martin Wechselberger
Vivien Kirk • James Sneyd

Mathematical Analysis of Complex Cellular Activity

Review 1: Richard Bertram, Joël Tabak, Wondimu Teka, Theodore Vo, Martin Wechselberger: Geometric Singular Perturbation Analysis of Bursting Oscillations in Pituitary Cells

Review 2: Vivien Kirk, James Sneyd: The Nonlinear Dynamics of Calcium

 Springer

Richard Bertram
Department of Mathematics
Florida State University
Tallahasse, FL, USA

Joël Tabak
Department of Mathematics
Florida State University
Tallahassee, FL, USA

Wondimu Teka
Department of Mathematics
Indiana University – Purdue
 University Indianapolis
Indianapolis, IN, USA

Theodore Vo
Department of Mathematics and Statistics
Boston University
Boston, MA, USA

Vivien Kirk
Deparment of Mathematics
The University of Auckland
Auckland, New Zealand

Martin Wechselberger
Department of Mathematics
University of Sydney
Sydney, NSW, Australia

James Sneyd
Department of Mathematics
University of Auckland
Auckland, New Zealand

ISSN 2364-4532   ISSN 2364-4931 (electronic)
Frontiers in Applied Dynamical Systems: Reviews and Tutorials
ISBN 978-3-319-18113-4   ISBN 978-3-319-18114-1 (eBook)
DOI 10.1007/978-3-319-18114-1

Library of Congress Control Number: 2015953625

Mathematics Subject Classification (2010): 92C05, 92C30, 92C37, 34C15, 37G15, 37N25

Springer Cham Heidelberg New York Dordrecht London

Printed on acid-free paper

Springer International Publishing AG Switzerland is part of Springer Science+Business Media (www.
springer.com)

Preface to the Series

The subject of dynamical systems has matured over a period of more than a century. It began with Poincare's investigation into the motion of the celestial bodies, and he pioneered a new direction by looking at the equations of motion from a qualitative viewpoint. For different motivation, statistical physics was being developed and had led to the idea of ergodic motion. Together, these presaged an area that was to have significant impact on both pure and applied mathematics. This perspective of dynamical systems was refined and developed in the second half of the twentieth century and now provides a commonly accepted way of channeling mathematical ideas into applications. These applications now reach from biology and social behavior to optics and microphysics.

There is still a lot we do not understand and the mathematical area of dynamical systems remains vibrant. This is particularly true as researchers come to grips with spatially distributed systems and those affected by stochastic effects that interact with complex deterministic dynamics. Much of current progress is being driven by questions that come from the applications of dynamical systems. To truly appreciate and engage in this work then requires us to understand more than just the mathematical theory of the subject. But to invest the time it takes to learn a new sub-area of applied dynamics without a guide is often impossible. This is especially true if the reach of its novelty extends from new mathematical ideas to the motivating questions and issues of the domain science.

It was from this challenge facing us that the idea for the *Frontiers in Applied Dynamics* was born. Our hope is that through the editions of this series, both new and seasoned dynamicists will be able to get into the applied areas that are defining modern dynamical systems. Each article will expose an area of current interest and excitement, and provide a portal for learning and entering the area. Occasionally we will combine more than one paper in a volume if we see a related audience as we have done in the first few volumes. Any given paper may contain new ideas and results. But more importantly, the papers will provide a survey of recent activity and the necessary background to understand its significance, open questions and mathematical challenges.

Editors-in-Chief
Christopher K R T Jones, Björn Sandstede, Lai-Sang Young

Preface

In the world of cell biology, there is a myriad of oscillatory processes, with periods ranging from the day of a circadian rhythm to the milliseconds of a neuronal action potential. To one extent or another they all interact, mostly in ways that we do not understand at all, and for at least the past 70 years, they have provided a fertile ground for the joint investigations of theoreticians and experimentalists. Experimentalists study them because they are physiologically important, while theoreticians tend to study them, not only for this reason, but also because such complex dynamic processes provide an opportunity to use, as their tools of investigation, the methods of mathematical analysis.

In this volume, we are concerned with two of these oscillatory processes: calcium oscillations and bursting electrical oscillations. These two are not chosen at random. Not only have they both been studied in depth by modellers and mathematicians, but we also have a good understanding – although not a complete one – of how they interact, and how one oscillatory process affects the other. They thus make an excellent example of how multiple oscillatory processes interact within a cell, and how mathematical methods can be used to understand such interactions better.

The theoretical study of electrical oscillations in cells began, to all intents and purposes, with the classic work of Hodgkin and Huxley in the 1950s. In a famous series of papers they showed how action potentials in neurons arose from the time-dependent control of the conductance of Na^+ and K^+ channels. The model they wrote down, a system of four coupled nonlinear ordinary differential equations, became one of the most influential models in all of physiology. It was quickly taken up by other modellers, who extended the model to study oscillations of electric potential in neurons, and over the last few decades the theoretical study of neurons and groups of neurons has expanded to become one of the largest and most active areas in all of mathematical biology.

More traditional applied mathematicians were also strongly influenced, albeit at one remove, by the Hodgkin-Huxley equations. The simplification by FitzHugh in the 1960s led to the FitzHugh-Nagumo model of excitability (Nagumo, a Japanese engineer, derived the same equation independently at the same time, from entirely different first principles) which formed the basis of more theoretical studies of excitability across many different areas, both inside and outside cell biology.

Oscillations in the cytosolic concentration of free Ca^{2+} (usually simply called Ca^{2+} oscillations) have a more recent history, not having been discovered until the development of Ca^{2+} fluorescent dyes in the 1980s allowed the measurement of intracellular Ca^{2+} concentrations with enough temporal precision. But since then, the number of theoretical and experimental investigations of Ca^{2+} oscillations has expanded rapidly. Calcium oscillations are now known to control a wide variety of cellular functions, including muscular contraction, water transport, gene differentiation, enzyme and neurotransmitter secretion, and cell differentiation. Indeed, the more we learn about intracellular Ca^{2+}, the more we realize how important it is for cellular function. Conversely, the intricate spatial and temporal behaviors exhibited by the intracellular Ca^{2+} concentration, including periodic plane waves, spiral waves, complex whole-cell oscillations, phase waves, stochastic resonance, and spiking, have encouraged theoreticians to use their skills, in collaboration with the experimentalists, to try and understand the dynamics of this ubiquitous ion.

Many cell types, however, contain both a membrane oscillator and a Ca^{2+} oscillator. The best-known examples of this, and the most widely studied, are the neuroendocrine cells of the hypothalamus and pituitary, as well as the endocrine cells of the pancreas, the pancreatic β cells. In these cell types, membrane oscillators and calcium oscillators are indissolubly linked; fast oscillations of the membrane potential open voltage-gated Ca^{2+} channels which allow Ca^{2+} to flow into the cell, which in turn activates the exocytotic machinery to secrete insulin (in the case of pancreatic β cells) or a variety of hormones (in the case of hypothalamic and pituitary cells). However, in each of these cell types, cytosolic Ca^{2+} also controls the conductance of membrane ion channels, particularly Ca^{2+}-sensitive K^+ and Cl^- channels, which in turn affect the membrane potential oscillations. In these endocrine cells, it is thus necessary to understand both types of cellular oscillator in order to understand overall cellular behavior.

Thus, this current volume. In it we first see how the interaction of Ca^{2+} cytosolic with membrane ion channels can result in the complex patterns of electrical spiking that we see in cells. We then discuss the basic theory of Ca^{2+} oscillations (common to almost all cell types), including spatio-temporal behaviors such as waves, and then review some of the theory of mathematical models of electrical bursting pituitary cells.

Although our understanding of how cellular oscillators interact remains rudimentary at best, this coupled oscillator system has been instrumental in developing our understanding of how the cytosol interacts with the membrane to form complex electrical firing patterns. In addition, from the theoretical point of view it has provided the motivation for the development and use of a wide range of mathematical methods, including geometric singular perturbation theory, nonlinear bifurcation theory, and multiple-time-scale analysis.

It is thus an excellent example of how mathematics and physiology can learn from each other, and work jointly towards a better understanding of complex cellular processes.

Tallahasse, FL, USA	Richard Bertram
Auckland, New Zealand	Vivien Kirk
Auckland, New Zealand	James Sneyd
Tallahasse, FL, USA	Joël Tabak
Indianapolis, IN, USA	Wondimu Teka
Boston, MA, USA	Theodore Vo
Sydney, NSW, Australia	Martin Wechselberger

Contents

Chapter 1
Geometric Singular Perturbation Analysis of Bursting Oscillations in Pituitary Cells

Richard Bertram, Joël Tabak, Wondimu Teka, Theodore Vo, and Martin Wechselberger

Abstract Dynamical systems theory provides a number of powerful tools for analyzing biological models, providing much more information than can be obtained from numerical simulation alone. In this chapter, we demonstrate how geometric singular perturbation analysis can be used to understand the dynamics of bursting in endocrine pituitary cells. This analysis technique, often called "fast/slow analysis," takes advantage of the different time scales of the system of ordinary differential equations and formally separates it into fast and slow subsystems. A standard fast/slow analysis, with a single slow variable, is used to understand bursting in pituitary gonadotrophs. The bursting produced by pituitary lactotrophs, somatotrophs, and corticotrophs is more exotic, and requires a fast/slow analysis with two slow variables. It makes use of concepts such as canards, folded singularities, and mixed-mode oscillations. Although applied here to pituitary cells, the approach can and has been used to study mixed-mode oscillations in other systems, including neurons, intracellular calcium dynamics, and chemical systems. The electrical bursting pattern produced in pituitary cells differs fundamentally from bursting oscillations

R. Bertram (✉)
Department of Mathematics, Florida State University, Tallahassee, FL, USA
e-mail: bertram@math.fsu.edu;

J. Tabak
Department of Mathematics and Biological Science, Florida State University,
Tallahassee, FL, USA
e-mail: joel@neuro.fsu.edu

W. Teka
Department of Mathematics, Indiana University – Purdue University Indianapolis, Indianapolis,
IN, USA
e-mail: wondimuwub@gmail.com

T. Vo
Department of Mathematics and Statistics, Boston University, Boston, MA
e-mail: theovo@bu.edu

M. Wechselberger
Department of Mathematics, University of Sydney, Sydney, NSW, Australia
e-mail: wm@maths.usyd.edu.au

© Springer International Publishing Switzerland 2015
R. Bertram et al., *Mathematical Analysis of Complex Cellular Activity*, Frontiers
in Applied Dynamical Systems: Reviews and Tutorials 1,
DOI 10.1007/978-3-319-18114-1_1

in neurons, and an understanding of the dynamics requires very different tools from those employed previously in the investigation of neuronal bursting. The chapter thus serves both as a case study for the application of recently-developed tools in geometric singular perturbation theory to an application in biology and a tutorial on how to use the tools.

1 Introduction

Techniques from dynamical systems theory have long been utilized to understand models of excitable systems, such as neurons, cardiac and other muscle cells, and many endocrine cells. The seminal model for action potential generation was published by Hodgkin and Huxley in 1952 and provided an understanding of the *biophysical basis* of electrical excitability (Hodgkin and Huxley (1952)). A mathematical understanding of the *dynamic mechanism* underlying excitability was provided nearly a decade later by the work of Richard FitzHugh (FitzHugh (1961)). He developed a planar model that exhibited excitability, and that could be understood in terms of phase plane analysis. A subsequent planar model, published in 1981 by Morris and Lecar, introduced biophysical aspects into the planar framework by incorporating ionic currents into the model, making the Morris-Lecar model a very useful hybrid of the four-dimensional biophysical Hodgkin-Huxley model and the two-dimensional mathematical FitzHugh model (Morris and Lecar (1981)). These planar models serve a very useful purpose: they allow one to use powerful mathematical tools to understand the dynamics underlying a biological phenomenon.

In this chapter, we use a similar approach to understand the dynamics underlying a type of electrical pattern often seen in endocrine cells of the pituitary. This pattern is more complex than the activity patterns studied by FitzHugh, and to understand it we employ dynamical systems techniques that did not exist when FitzHugh did his groundbreaking work. Indeed, the mathematical tools that we employ, geometric singular perturbation analysis with a focus on folded singularities, are still being developed (Brons et al. (2006), Desroches et al. (2008a), Fenichel (1979), Guckenheimer and Haiduc (2005), Szmolyan and Wechselberger (2001; 2004), Wechselberger (2005; 2012)). The techniques are appealing from a purely mathematical viewpoint (see Desroches et al. (2012) for review), but have also been used in applications. In particular, they have been employed successfully in the field of neuroscience (Erchova and McGonigle (2008), Rubin and Wechselberger (2007; 2008), Wechselberger and Weckesser (2009)), intracellular calcium dynamics (Harvey et al. (2010; 2011)), and chemical systems (Guckenheimer and Scheper (2011)). As we demonstrate in this chapter, these tools are also very useful in the analysis of the electrical activity of endocrine pituitary cells. We emphasize, however, that the analysis techniques can and have been used in many other settings, so this chapter can be considered a case study for biological application, as well as a tutorial on how to perform a geometric singular perturbation analysis of a system with multiple time scales.

The anterior region of the pituitary gland contains five types of endocrine cells that secrete a variety of hormones, such as prolactin, growth hormone, and luteinizing hormone, into the blood. These pituitary hormones are transported by the vasculature to other regions of the body where they act on other endocrine glands, which in turn secrete their hormones into the blood, and on other tissue including the brain. The pituitary gland thus acts as a master gland. Yet the pituitary does not act independently, but instead is controlled by neurohormones released from neurons of the hypothalamus, which is located nearby.

Many endocrine cells, including anterior pituitary cells, release hormones through a *stimulus-secretion coupling* mechanism. When the cell receives a stimulatory message, there is an increase in the concentration of intracellular Ca^{2+} that triggers the hormone secretion. More often than not, the response to the input is a rhythmic output due to oscillations in the Ca^{2+} concentration. Here we are interested in the dynamics of these Ca^{2+} oscillations. There are actually two possibilities, and both can be found in pituitary cells. First, Ca^{2+} oscillations can be due to the cell's electrical activity. In this case, oscillations in electrical activity bring Ca^{2+} into the cell through ion channels in the plasma membrane. This is called a plasma membrane oscillator, because the channels responsible for electrical activity and letting in Ca^{2+} are on the cell membrane. Another mechanism for intracellular Ca^{2+} oscillations is the periodic release of Ca^{2+} from intracellular stores, through channels on the membrane of these stores. The main Ca^{2+}-storing organelle is the endoplasmic reticulum (ER), so this mechanism is called an ER oscillator. In both cases we get rhythmic Ca^{2+} increases. Although the two mechanisms can interact, we will not look deeply into their interactions here and instead focus on each separately. This chapter describes work performed to understand the dynamics underlying these two types of rhythmic Ca^{2+} increase that underlie hormone secretion from the endocrine cells of the anterior pituitary.

Like neurons and other excitable cells, pituitary cells can generate brief electrical impulses (also called action potentials or spikes). Different ion concentrations across the plasma membrane and ion channels specific for certain types of ions create a difference in the electrical potential across the membrane (the *membrane potential*, *V*). Electrical activity in the form of impulses is caused by the regenerative opening of membrane ion channels, which allows ions through the membrane according to their concentration gradient. The opening of channels is controlled by *V*, which accounts for positive and negative feedback mechanisms. Usually channels open when *V* increases (*depolarizes*), so channels permeable to Na^+ or Ca^{2+} which flows into the cell and thus creates inward currents that further depolarize the membrane, will provide the positive feedback that underlies the rapid rise of *V* at the beginning of a spike. Channels permeable to K^+, which is more concentrated inside the cells, produce an outward current that acts as negative feedback to decrease *V* and to terminate a spike. There are many types of ion channels expressed in pituitary cells, and the combination of ionic currents mediated by these channels determines the pattern of spontaneous electrical activity exhibited by the cells (see Stojilković et al. (2010) for review). In a physiological setting, this spontaneous activity is subject

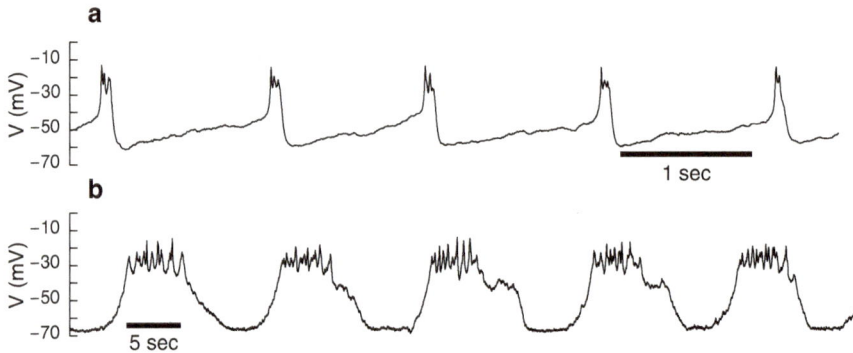

Fig. 1 Recordings of electrical bursting using the perforated patch method with amphotericin B. (**A**) Bursting in an unstimulated cell from the GH4C1 lacto-somatotroph cell line. (**B**) Bursting in a pituitary gonadotroph stimulated with GnRH (1 nM). Note the different time scale

to continuous adjustment by hypothalamic neuropeptides, by hormones from other glands such as the testes or ovaries, and by other pituitary hormones (Freeman (2006), Stojilković et al. (2010)).

One typical pattern of electrical activity in pituitary cells is bursting. This consists of episodes of spiking followed by quiescent phases, repeated periodically. Such bursting oscillations have been observed in the spontaneous activity of prolactin-secreting lactotrophs, growth hormone-secreting somatotrophs, and ACTH-secreting corticotrophs (Van Goor et al. (2001a;b), Kuryshev et al. (1996), Tsaneva-Atanasova et al. (2007)), as well as GH4C1 lacto-somatotroph tumor cells (Tabak et al. (2011)). The bursting pattern has a short period and the spikes tend to be very small compared with those of tonically spiking cells (Fig. 1A). In fact, the spikes don't look much like impulses at all, but instead appear more like small oscillations. This type of bursting is often referred to as *pseudo-plateau bursting* (Stern et al. (2008)). A very different form of bursting is common in gonadotrophs that have been stimulated by gonadotropin releasing hormone (GnRH), their primary activator (Li et al. (1995; 1994), Tse and Hille (1992)), as well as other stimulating factors (Stojilković et al. (2010)). These bursts have much longer period than the spontaneous pseudo-plateau bursts (Fig. 1B). Since the biophysical basis for this bursting pattern is periodic release of Ca^{2+} from an internal store, we refer to it as *store-generated bursting*. Both forms of bursting elevate the Ca^{2+} concentration in the cytosol of the cell and evoke a higher level of hormone secretion than do tonic spiking patterns (Van Goor et al. (2001b)). This is the main reason that endocrinologists are interested in electrical bursting in pituitary cells, which in turn motivates mathematicians to develop and analyze models of the cells' electrical activity.

Bursting patterns also occur in neurons (Crunelli et al. (1987), Del Negro et al. (1998), Lyons et al. (2010), Nunemaker et al. (2001)) and in pancreatic β-cells, another type of endocrine cell that secretes the hormone insulin (Dean and Mathews (1970), Bertram et al. (2010)). The ubiquity of the oscillatory pattern and its complexity has attracted a great deal of attention from mathematicians, who have used various techniques to study the mechanism(s) underlying the bursting pattern. The earliest models of bursting neurons were developed in the 1970s, and bursting models have been published regularly ever since. Over the past decade several books have described some of these models and the techniques used to analyze them (Coombes and Bressloff (2005), Izhikevich (2007), Keener and Sneyd (2008)). The primary analysis technique takes advantage of the difference in time scales between variables that change quickly and those that change slowly. This "fast/slow analysis" or "geometric singular perturbation analysis" was pioneered by John Rinzel in the 1980s (Rinzel (1987)) and has been extended in subsequent years (Coombes and Bressloff (2005)). While modeling and analysis of bursting in neurons and pancreatic β-cells has a long history and is now well developed, the construction and analysis of models of bursting in pituitary cells is at a relatively early stage. The burst patterns in pituitary cells are very different from those in cells studied previously, and the fast/slow analysis technique used in neurons is of limited use for studying pseudo-plateau bursting in pituitary cells (Toporikova et al. (2008), Teka et al. (2011a)). Instead, a new fast/slow analysis technique has been developed for pseudo-plateau bursting that relies on concepts such as folded singularities, canards, and the theory of mixed-mode oscillations (Teka et al. (2011a), Vo et al. (2010)). In the first part of this chapter we describe this technique and how it relates to the original fast/slow analysis technique used to analyze other cell types.

One fundamental difference between the spontaneous bursting observed in many lactotrophs and somatotrophs and that seen in stimulated gonadotrophs is that in the former the periodic elevations of intracellular Ca^{2+} are in phase with the electrical activity, while in the latter they are 180° out of phase. This is because the former is driven by electrical activity, which brings Ca^{2+} into the cell through plasma membrane ion channels, while the latter is driven by the ER oscillator, which periodically releases a flood of Ca^{2+} into the cytosol. This Ca^{2+} binds to Ca^{2+}-activated K^+ channels and activates them, resulting in a lowering (*hyperpolarization*) of the membrane potential and terminating the spiking activity. Thus, each time that the Ca^{2+} concentration is high it turns off the electrical activity. In the second part of this chapter we describe a model for this store-operated bursting and demonstrate how it can be understood in terms of coupled electrical and Ca^{2+} oscillators, again making use of fast/slow analysis.

2   The Lactotroph/Somatotroph Model

We use a model for the pituitary lactotroph developed in Tabak et al. (2007) and recently used in Teka et al. (2011b), Teka et al. (2011a, 2012), and Tomaiuolo et al. (2012). This model can also be thought of as a model for the pituitary somatotroph, since lactotrophs and somatotrophs exhibit similar behaviors and the level of detail in the model is insufficient to distinguish the two. This consists of ordinary differential equations for the membrane potential or voltage (V), an activation variable describing the fraction of activated K$^+$ channels (n), and the intracellular free Ca^{2+} concentration (c):

$$C_m \frac{dV}{dt} = -[I_{Ca}(V) + I_K(V,n) + I_{SK}(V,c) + I_{BK}(V)] \tag{1.1}$$

$$\frac{dn}{dt} = \frac{n_\infty(V) - n}{\tau_n} \tag{1.2}$$

$$\frac{dc}{dt} = -f_c(\alpha I_{Ca} + k_c c). \tag{1.3}$$

The parameter C_m in Eq. 1.1 is the membrane capacitance, and the right-hand side is the sum of ionic currents. I_{Ca} is an inward current carried by Ca^{2+} flowing through Ca^{2+} channels and is responsible for the upstroke of an action potential. It is assumed to activate instantaneously, so no activation variable is needed. The current is

$$I_{Ca}(V) = g_{Ca} m_\infty(V)(V - V_{Ca}) \tag{1.4}$$

where g_{Ca} is the maximum conductance (a parameter) and the instantaneous activation of the current is described by

$$m_\infty(V) = \left(1 + \exp\left(\frac{v_m - V}{s_m}\right)\right)^{-1}. \tag{1.5}$$

The parameters v_m and s_m set the half-maximum location and the slope, respectively, of the Boltzman curve. Since this is an increasing function of V, I_{Ca} becomes activated as V increases from its low resting value toward v_m. The driving force for the current is $(V - V_{Ca})$, where V_{Ca} is the Nernst potential for Ca^{2+}.

I_K is an outward delayed-rectifying K$^+$ current with activation that is slower than that for I_{Ca}. This current, largely responsible for the downstroke of a spike, is

$$I_K(V,n) = g_K n(V - V_K) \tag{1.6}$$

where g_K is the maximum conductance, V_K is the K$^+$ Nernst potential, and the activation of the current is described by Eq. 1.2. The steady state activation function for n is

$$n_\infty(V) = \left(1 + \exp\left(\frac{v_n - V}{s_n}\right)\right)^{-1} \tag{1.7}$$

and the rate of change of n is determined by the time constant τ_n.

Some K$^+$ channels are activated by intracellular Ca^{2+}, rather than by voltage. One type of Ca^{2+}-activated K$^+$ channel is the SK channel (small conductance K(Ca) channel). Because channel activation is due to the accumulation of Ca^{2+} in the cell (i.e., an increase in c), and this occurs more slowly than changes in V, the current through SK channels contributes little to the spike dynamics. Instead, it contributes to the patterning of spikes. The current through this channel is modeled here by

$$I_{SK}(V, c) = g_{SK} s_{\infty}(c)(V - V_K) \tag{1.8}$$

where g_{SK} is the maximum conductance and the c-dependent activation function is

$$s_{\infty}(c) = \frac{c^2}{c^2 + K_d^2} \tag{1.9}$$

where K_d is the Ca^{2+} level of half activation.

The final current in the model reflects K$^+$ flow through other Ca^{2+}-activated K$^+$ channels called BK channels (large conductance K(Ca) channels). These channels are located near Ca^{2+} channels and are gated by V and by the high-concentration Ca^{2+} nanodomains that form at the mouth of the open channel. As has been pointed out previously (Sherman et al. (1990)), the Ca^{2+} seen by the BK channel reflects the state of the Ca^{2+} channel, which is determined by the membrane potential. Thus, activation of the BK current can be modeled as a V-dependent process:

$$I_{BK}(V) = g_{BK} b_{\infty}(V)(V - V_K) \tag{1.10}$$

where

$$b_{\infty}(V) = \left(1 + \exp\left(\frac{v_b - V}{s_b}\right)\right)^{-1}. \tag{1.11}$$

Because this current activates rapidly with changes in voltage (due to the rapid formation of Ca^{2+} nanodomains), it limits the upstroke and contributes to the downstroke of an action potential.

The differential equation for the free intracellular Ca^{2+} concentration (Eq. 1.3) describes the influx of Ca^{2+} into the cell through Ca^{2+} channels (αI_{Ca}) and the efflux through Ca^{2+} pumps $k_c c$. The parameter α converts current to molar flux and the parameter k_c is the pump rate. Finally, parameter f_c is the fraction of Ca^{2+} in the cell that is free, i.e., not bound to Ca^{2+} buffers. Default values of all parameters are listed in Table 1.

Table 1 Default parameter values for the lactotroph model

$g_{Ca} = 2\,\text{nS}$	$g_K = 4\,\text{nS}$	$g_{SK} = 1.7\,\text{nS}$	$g_{BK} = 0.4\,\text{nS}$
$V_{Ca} = 50\,\text{mV}$	$V_K = -75\,\text{mV}$	$C_m = 10\,\text{pF}$	$\alpha = 1.5 \times 10^{-3}\,\text{pA}^{-1}\mu\text{M}$
$\tau_n = 43\,\text{ms}$	$f_c = 0.01$	$k_c = 0.16\,\text{ms}^{-1}$	$K_d = 0.5\,\mu\text{M}$
$v_n = -5\,\text{mV}$	$s_n = 10\,\text{mV}$	$v_m = -20\,\text{mV}$	$s_m = 12\,\text{mV}$
$v_b = -20\,\text{mV}$	$s_b = 5.6\,\text{mV}$		

3   The Standard Fast/Slow Analysis

The three model variables change on different time scales. The time constant for the membrane potential is the product of the capacitance and the input resistance: $\tau_V = C_m/g_{total}$, where $g_{total} = g_{Ca} + g_K + g_{SK} + g_{BK}$ is the total membrane conductance. This varies with time as V changes, and during the burst shown in Fig. 2, g_{total} ranges from about 0.5 nS during the silent phase of the burst to about 3 nS during the active phase of the burst, so $3.3 < \tau_V < 20$ mS. The variable n has a time constant of $\tau_n = 43$ ms. The time constant for c is $\frac{1}{f_c k_c} = 625$ ms. Hence, $\tau_V < \tau_n < \tau_c$ and V is the fastest variable, while c is the slowest.

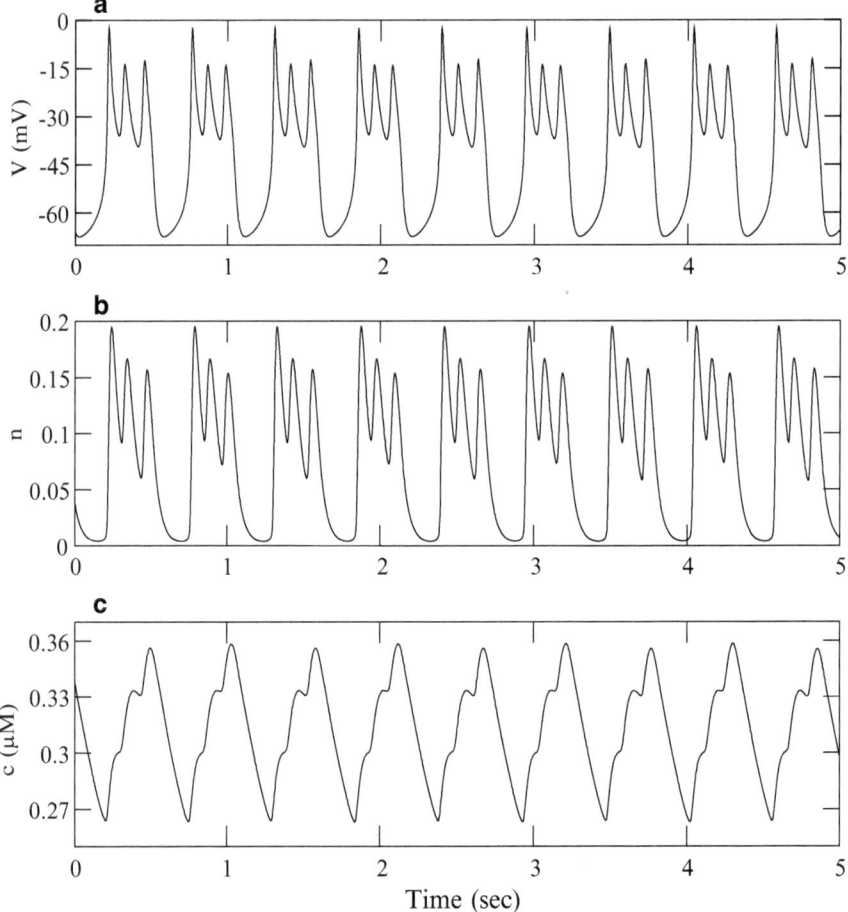

Fig. 2 Bursting produced by the lactotroph model. (**A**) Voltage V exhibits small spikes emerging from a plateau. (**B**) The variable n is sufficiently fast to reliably follow V. (**C**) The variable c changes on a much slower time scale, exhibiting a saw-tooth time course

The time courses of the three variables shown in Fig. 2 confirm the differences in time scales. The spikes that occur during each burst in V are reliably reflected in n, but are dampened in c. Indeed, c is an accumulating variable, similar to what one observes in the recovery variable during a relaxation oscillation. This observation motivates the idea of analyzing the burst trajectory just as one would analyze a relaxation oscillation with a fast variable V and a slow recovery variable c. That is, the trajectory is examined in the c-V plane and the c and V nullclines are utilized. However, since the system is 3-dimensional, one replaces the nullcline of the fast variable (V) with the fast subsystem (V and n) bifurcation diagram, where the slow variable c is treated as the bifurcation parameter. This is the fundamental idea of the standard fast/slow analysis, which is illustrated in Fig. 3A. The fast-subsystem bifurcation diagram, often called the z-curve, consists of a bottom branch of stable

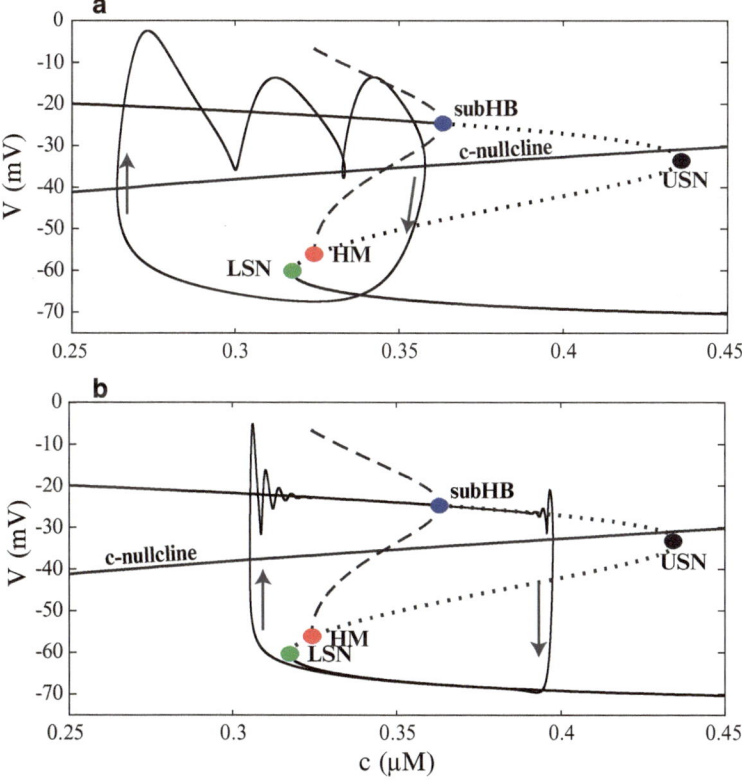

Fig. 3 2-fast/1-slow analysis of pseudo-plateau bursting. The 3-branched z-curve consists of stable (solid) and unstable (dotted) equilibria and a branch of unstable periodic solutions (dashed). Bifurcations include a lower saddle-node (LSN), upper saddle-node (USN), subcritical Hopf (subHB), and homoclinic (HM) bifurcations. (**A**) With default parameter values, the burst trajectory (thick black curve) only partially follows the z-curve. (**B**) When the slow variable is made slower by reducing f_c from 0.01 to 0.001 the full-system trajectory follows the z-curve much more closely

steady states (solid curve), a middle branch of unstable saddle points (dotted curve), and a top branch of stable and unstable steady states. The three branches are joined by lower and upper saddle-node bifurcations (LSN and USN, respectively), and the stability of the top branch changes at a subcritical Hopf bifurcation (subHB). The Hopf bifurcation gives rise to a branch of unstable periodic solutions that terminates at a homoclinic bifurcation (HM). Thus, we see that the fast subsystem has an interval of c values where it is bistable between lower (hyperpolarized) and upper (depolarized) steady states. This interval extends from LSN to subHB. The c nullcline intersects the z-curve between subHB and USN. This intersection is an unstable equilibrium of the full system of equations.

The next step in the fast/slow analysis is to superimpose the burst trajectory and analyze the dynamics using a phase plane approach. Since the c variable is much slower than V, the trajectory largely follows the z-curve, as it would follow the nullcline of the fast variable during a relaxation oscillation. Below the c-nullcline the flow is to the left, and above the nullcline it is to the right. Hence, during the silent phase of the burst the trajectory moves leftward along the bottom branch of the z-curve. When LSN is reached there is a fast jump up to the top branch of the z-curve. The trajectory follows this rightward until subHB is reached, at which point it jumps down to the bottom branch of the z-curve, restarting the cycle.

As is clear from Fig. 3A, the trajectory does not follow the z-curve very closely. One explanation for this is that the equilibria on the top branch are weakly attracting foci, and the "slow variable" c changes too quickly for the trajectory to ever get close to the branch of foci. Thus, weakly damped oscillations are produced during the active phase, and these damped oscillations are the spikes of the burst. This interpretation is supported in Fig. 3B, where the slow variable is made 10-times slower by decreasing f_c from 0.01 to 0.001. Now the trajectory moves much more closely along both branches of the z-curve. During the active phase there are a few initial oscillations which quickly dampen. Once the trajectory passes through subHB there is a slow passage effect (Baer et al. (1989), Baer and Gaekel (2008)) and a few growing oscillations before the trajectory jumps down to the lower branch.

This analysis, which we will call a *2-fast/1-slow analysis*, provides some useful information about the bursting. For example, this approach was used to understand the mechanism for active phase termination during a burst, by constructing the 2-dimensional stable manifold of the fast subsystem saddle point (Nowacki et al. (2010)). This approach was also used to understand the complex burst resetting that occurs in response to upward voltage perturbations (Stern et al. (2008)). We have shown how the z-curve for this pseudo-plateau bursting relates to that for the *plateau bursting* often observed in neurons (Teka et al. (2011a)). This is illustrated in Fig. 4, using the Chay-Keizer model for bursting in pancreatic β-cells (Chay and Keizer (1983)). (The equations for this model are given in the Appendix.) The standard z-curve for plateau bursting is shown in panel A. It is characterized by a branch of stable periodic solutions that are the spikes of the burst. In this figure they emanate from a supercritical Hopf bifurcation (supHB). With this stable periodic branch, the spikes tend to be much larger than those produced during pseudo-plateau bursting and they do not dampen as the active phase progresses. If the activation curve

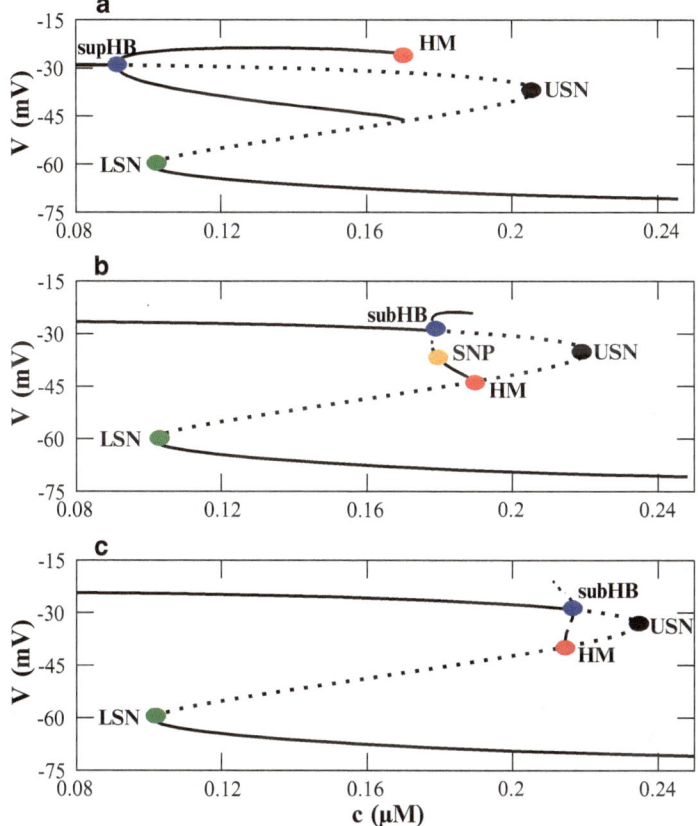

Fig. 4 The Chay-Keizer model is used to illustrate the transition between plateau and pseudo-plateau bursting. (**A**) The z-curve for plateau bursting, using default parameter values given in Appendix, is characterized by a branch of stable periodic spiking solutions arising from a supercritical Hopf bifurcation (supHB). (**B**) Increasing the value of v_n from -16 mV to -14 mV moves the Hopf bifurcation rightward and converts it to a subHB, with an associated saddle-node of periodics (SNP) bifurcation. (**C**) Increasing v_n further to -12 mV creates the z-curve that characterizes pseudo-plateau bursting. From Teka et al. (2011b)

for the hyperpolarizing K^+ current is moved rightward by increasing v_n, the cell becomes more excitable. As a result, the Hopf bifurcation moves rightward and becomes subcritical (Fig. 4B). Most importantly, the region of bistability between a stable spiking solution and a stable hyperpolarized steady state has largely been replaced by bistability between two stable steady states of the fast subsystem: one hyperpolarized and one depolarized. When the activation curve is shifted further to the right (Fig. 4C), the stable periodic branch has been entirely replaced by a stable stationary branch and the z-curve is that for pseudo-plateau bursting. Other maneuvers that make the cell more excitable, such as moving the activation curve for the depolarizing I_{Ca} current leftward, increasing the conductance g_{Ca} for this current or decreasing the conductance g_K for the hyperpolarizing I_K current, have the same effect on the z-curve (Teka et al. (2011b)). In addition to changing the fast-subsystem bifurcation diagram, the speed of the slow variable must also be modified to convert between plateau and pseudo-plateau bursting (it must be faster for pseudo-plateau bursting, which is achieved by increasing the value of f_c). In a separate study, Osinga and colleagues demonstrated that the fast-subsystem bifurcation structure of both plateau and pseudo-plateau bursting could be obtained by unfolding a codimension-4 bifurcation (Osinga et al. (2012)). This explains why the pseudo-plateau bifurcation structure was not seen in an earlier classification of bursting that was based on the unfolding of a codimension-3 bifurcation (Bertram et al. (1995)).

Although the 2-fast/1-slow analysis provides useful information about the pseudo-plateau bursting, it has some major shortcomings. Most obviously, the burst trajectory does not follow the z-curve very closely unless the slow variable is slowed down to the point where spikes no longer occur during the active phase (Fig. 3). Also, the explanation for the origin of the spikes is not totally convincing, since it is based on a local analysis of the steady states of the top branch, while the bursting trajectory is not near these steady states. It also provides no information about how many spikes to expect during a burst. Finally, as illustrated in Fig. 5, it fails to explain the transition that occurs from pseudo-plateau bursting to continuous spiking when the c-nullcline is lowered. In this figure, reducing the k_c parameter lowers the nullclline without affecting the z-curve. In both panels B and D the

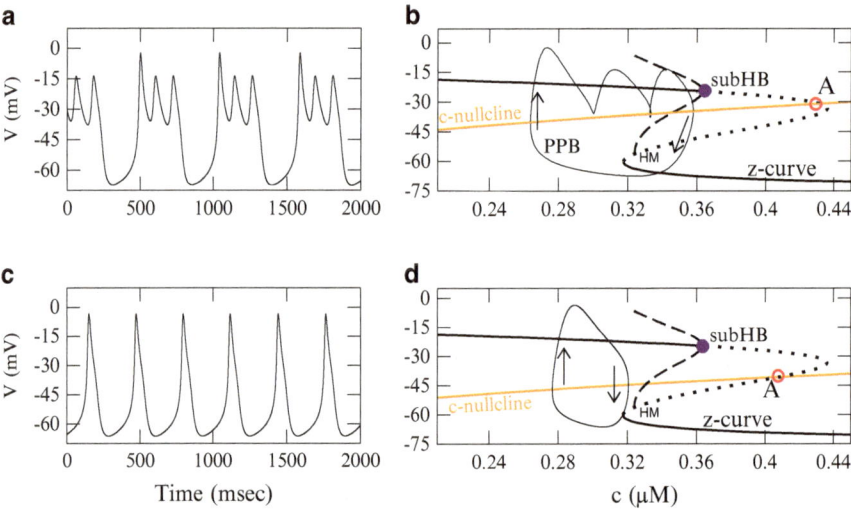

Fig. 5 A 2-fast/1-slow analysis fails to explain the transition from pseudo-plateau bursting to spiking in the lactotroph model when the c-nullcline is lowered. (**A**) Bursting produced using default parameter values. (**B**) The standard fast/slow analysis of the bursting pattern. (**C**) The bursting is converted to continuous spiking when k_c is reduced from 0.16 ms^{-1} to 0.1 ms^{-1}. (**D**) It is not apparent from the fast/slow analysis why the transition took place. From Teka et al. (2012)

nullcline intersects the z-curve to form an unstable full-system equilibrium (labeled as "A") as well as the unstable periodic branch, forming an unstable full-system periodic solution. Yet, in one case the system bursts (panel A), while in the other it spikes continuously (panel C). This is a clear indication that predictions made regarding pseudo-plateau bursting with this type of analysis may not be reliable.

4   The 1-Fast/2-Slow Analysis

In the analysis above, the variable with the intermediate time scale (n) was associated with the fast subsystem, and the bursting dynamics analyzed by comparing the full-system trajectory to what one would expect if the single slow variable (c) were very slow. That is, by going to the singular limit $f_c \to 0$ and constructing a fast-subsystem bifurcation diagram with c as the bifurcation parameter. Alternatively, one could associate n with the slow subsystem and then study the dynamics by comparing the bursting to what one would expect if the single fast variable were very fast. That is, by going to the singular limit $C_m \to 0$. We take this 1-fast/2-slow analysis approach here, where the variable V forms the fast subsystem and n and c form the slow subsystem. This is formalized using non-dimensional equations in Teka et al. (2011a) and Vo et al. (2010), where more details and derivations can also be found. A recent review of mixed-mode oscillations (Desroches et al. (2012)) gives more detail on the key dynamical structures described below.

4.1   Reduced, Desingularized, and Layer Systems

In the following, we assume that C_m is small, so that the V variable is in a pseudo-equilibrium state. Define the function f as the right-hand side of Eq. 1.1:

$$f(V, n, c) \equiv -(I_{Ca} + I_K + I_{SK} + I_{BK}). \tag{1.12}$$

and then

$$\tilde{f}(V, n, c) \equiv f(V, n, c)/g_{max} \tag{1.13}$$

where g_{max} is a representative conductance value, for example, the maximum conductance during an action potential. Then the dynamics of the fast subsystem are, in the singular limit, given by the *layer problem*:

$$\frac{dV}{dt_f} = \tilde{f}(V, n, c) \tag{1.14}$$

$$\frac{dn}{dt_f} = 0 \tag{1.15}$$

$$\frac{dc}{dt_f} = 0 \ . \tag{1.16}$$

where $t_f = (g_{max}/C_m)t$ is a dimensionless fast time variable. The equilibrium set of this subsystem is called the *critical manifold*, which is a surface in $\mathbb{R}^3$:

$$S \equiv \{(V, n, c) \in \mathbb{R}^3 : f(V, n, c) = 0\}. \tag{1.17}$$

Since f is linear in n, it is convenient to solve for n in terms of V and c:

$$n = n(V, c) = -\frac{1}{g_K}[h(V) + g_{SK}s_\infty(c)] \tag{1.18}$$

where

$$h(V) = g_{Ca}m_\infty(V)\left(\frac{V - V_{Ca}}{V - V_K}\right) + g_{BK}b_\infty(V). \tag{1.19}$$

The critical manifold is a folded surface consisting of three sheets connected by two fold curves (Fig. 6). The one-dimensional fast subsystem is bistable; for a range of values of n and c there is a stable hyperpolarized steady state and a stable depolarized steady state, separated by an unstable steady state. The stable steady states form the attracting lower and upper sheets of the critical manifold (denoted as S_a^+ and S_a^- and where $\frac{\partial f}{\partial V} < 0$), while the separating unstable steady states form the repelling middle sheet (denoted as S_r and where $\frac{\partial f}{\partial V} > 0$). The sheets are connected by fold curves denoted by L^+ and L^- that consist of points on the surface where

$$\frac{\partial f}{\partial V} = 0. \tag{1.20}$$

That is,

$$L^\pm \equiv \{(V, n, c) \in \mathbb{R}^3 : f(V, n, c) = 0 \text{ and } \frac{\partial f}{\partial V}(V, n, c) = 0\}. \tag{1.21}$$

The projection of the top fold curve onto the lower sheet is denoted $P(L^+)$, while the projection of the lower fold curve onto the top sheet is denoted $P(L^-)$. Both projections are shown in Fig. 6.

The critical manifold is not only the equilibrium set of the fast subsystem, but is also the phase space of the slow subsystem. This slow subsystem, also called the *reduced system*, is described by

$$f(V, n, c) = 0 \tag{1.22}$$

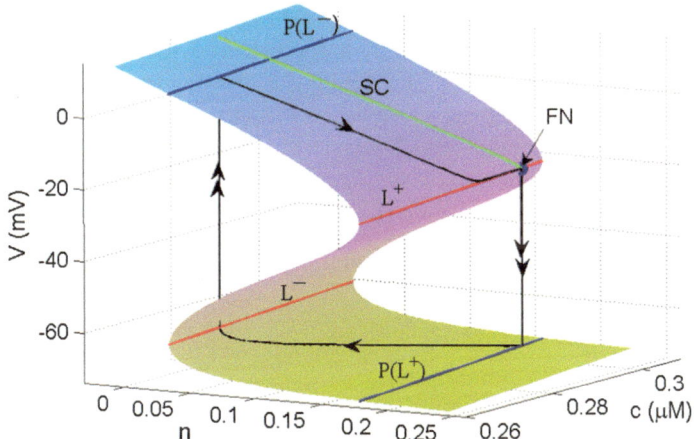

Fig. 6 The critical manifold is the set of points in $\mathbb{R}^3$ for which the fast variable V is at equilibrium (Eq. 1.17). The two fold curves are denoted by L^+ and L^-. The projections along the fast fibers of the fold curves are denoted by $P(L^+)$ and $P(L^-)$. Also shown is the folded node singularity (FN) and the strong canard (SC) that enters the folded node. From Teka et al. (2011a)

$$\frac{dn}{dt} = \frac{n_\infty(V) - n}{\tau_n} \tag{1.23}$$

$$\frac{dc}{dt} = -f_c(\alpha I_{Ca} + k_c c). \tag{1.24}$$

This differential-algebraic system describes the flow when the trajectory is on the critical manifold, which is given as a graph in Eq. 1.18. We can thus present the system in a single *coordinate chart* (V, c) including the neighborhood of the two folds. A condition is then needed to constrain the trajectories to the critical manifold. It is the total time derivative of $f = 0$ that provides this condition. That is,

$$\frac{d}{dt} f(V, n, c) = \frac{d}{dt} 0 \tag{1.25}$$

or

$$-\frac{\partial f}{\partial V}\frac{dV}{dt} = \frac{\partial f}{\partial c}\frac{dc}{dt} + \frac{\partial f}{\partial n}\frac{dn}{dt}. \tag{1.26}$$

Using Eqs 1.23, 1.24,

$$-\frac{\partial f}{\partial V}\frac{dV}{dt} = -f_c(\alpha I_{Ca} + k_c c)\frac{\partial f}{\partial c} + \left(\frac{n_\infty(V) - n(V, c)}{\tau_n}\right)\frac{\partial f}{\partial n}. \tag{1.27}$$

The reduced system then consists of the differential equations Eqs. 1.24 and 1.27 where $n(V, c)$ is given by Eq. 1.18.

The reduced system is singular at the fold curves (where $\frac{\partial f}{\partial V} = 0$), so the speed of a trajectory approaches ∞ as it approaches a fold curve. (This can be seen by solving Eq. 1.27 for $\frac{dV}{dt}$ and noting that the denominator approaches 0, but the numerator does not, as a fold curve is approached.) The singularity can be removed by introducing a rescaled time $d\tau = -(\frac{\partial f}{\partial V})^{-1} dt$. This produces a system that behaves like the reduced system, except at the fold curves, which are transformed into nullclines of the c variable. With this rescaled time, the following *desingularized system* is formed:

$$\frac{dV}{d\tau} = F(V, c) \tag{1.28}$$

$$\frac{dc}{d\tau} = f_c(\alpha I_{Ca} + k_c c)\frac{\partial f}{\partial V} \ , \tag{1.29}$$

where $F(V, c)$ is defined as

$$F(V, c) \equiv -f_c(\alpha I_{Ca} + k_c c)\frac{\partial f}{\partial c} + \left(\frac{n_\infty(V) - n(V, c)}{\tau_n}\right)\frac{\partial f}{\partial n}. \tag{1.30}$$

Like the reduced system, Eqs. 1.28–1.30 along with Eq. 1.18 describe the flow on the top and bottom sheets of the critical manifold. They also describe the flow on the middle sheet, but in this case the flow is backwards in time due to the time rescaling. The jump from one attracting sheet to another is described by the layer problem, which was discussed above.

A *singular periodic orbit* can be constructed by gluing together trajectories from the desingularized system and the layer system such that the resulting orbit returns to its starting point. An example is shown in Fig. 6. Beginning from a point on the singular periodic orbit that lies on S_a^+, the desingularized system is solved to yield a trajectory that moves along S_a^+ until it reaches L^+ (black curve with single arrow). From here, it moves to the bottom sheet following a fast fiber (black curve with double arrows). From a point on $P(L^+)$ the desingularized equations are again solved to yield a trajectory that moves along S_a^- until L^- is reached. The trajectory then moves along a fast fiber to a point on $P(L^-)$ on the top sheet. From here the desingularized equations are again solved and the trajectory continues until the starting point is reached.

4.2   Folded Singularities and the Origin of Pseudo-Plateau Bursting

There are two very different types of equilibria of the desingularized system: ordinary and folded singularities. An ordinary singularity of the desingularized system satisfies

$$f(V, n, c) = 0 \tag{1.31}$$

$$n = n_\infty(V) \tag{1.32}$$

$$c = c_\infty(V) = -f(\alpha I_{Ca} + k_c c) \tag{1.33}$$

and is an equilibrium of the full system Eqs. 1.1–1.3 . A folded singularity lies on a fold curve and satisfies

$$f(V, n, c) = 0 \tag{1.34}$$

$$F(V, c) = 0 \tag{1.35}$$

$$\frac{\partial f}{\partial V} = 0. \tag{1.36}$$

As previously noted, in the reduced system (Eqs. 1.24, 1.27, and 1.18), trajectories pass through a fold curve with infinite velocity. Folded singularities are an exception: at these points *both* numerator and denominator approach 0, and hence a trajectory passes through a folded singularity with finite speed. In the full system near the singular limit, the trajectory can pass through the fold curve and move along the middle sheet of the slow manifold for some time before jumping off.

A linear stability analysis of a folded singularity indicates whether it is a folded node (two real eigenvalues of the same sign), folded saddle (real eigenvalues of opposite sign), or folded focus (complex conjugate pair of eigenvalues). In the full system, *singular canards* exist in the neighborhood of a folded node and a folded saddle (Benoit (1983), Szmolyan and Wechselberger (2001)). These trajectories enter the folded singularity, in our case along S_a^+, and move through it in finite time, emerging on the repelling sheet S_r and traveling along this sheet for some time. For the parameter values used in Fig. 6 there is a folded node (FN) on L^+. In such a case, there is a whole sector of singular canards, bounded by L^+ and the *strong singular canard* (denoted by SC in Fig. 6) associated with the trajectory that is tangent to the eigendirection of the strong eigenvalue of the FN. This sector is called the *singular funnel*. A singular periodic orbit that enters the singular funnel will exhibit *canard-induced mixed-mode oscillations* (MMOs) away from the singular limit (i.e., when $C_m > 0$) (Brons et al. (2006)).

According to Fenichel theory (Fenichel (1979)), for $C_m > 0$ the critical manifold perturbs smoothly to a *slow manifold* consisting of invariant attracting and repelling manifolds. We denote the attracting manifolds as S_{a,C_m}^+ and S_{a,C_m}^-, and the repelling manifold as S_{r,C_m}. Since the critical manifold loses hyperbolicity at L^+ and L^-, Fenichel theory does not apply there. Indeed, the critical manifold near a folded node perturbs to twisted sheets (Guckenheimer and Haiduc (2005), Wechselberger (2005)). This is illustrated in Fig. 7, where S_{a,C_m}^+ (blue) and S_{r,C_m} (red) come together near the FN. The numerical technique used to compute the twisted sheets utilizes continuation of trajectories that satisfy boundary value problems, and was developed in Desroches et al. (2008a) and Desroches et al. (2008b).

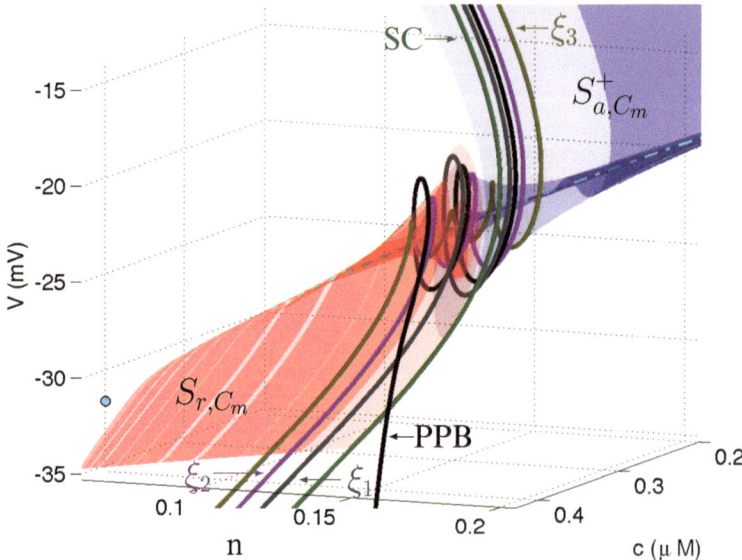

Fig. 7 The twisted slow manifold near a folded node, calculated using $C_m = 2$ pF with default parameter values. The primary strong canard (SC, green) flows from S^+_{a,C_m} to S_{r,C_m} with a half rotation. The secondary canard ξ_1 flows from S^+_{a,C_m} to S_{r,C_m} with a single rotation. The other secondary canards (ξ_2, ξ_3) have two and three rotations, respectively. The full system has an unstable equilibrium near S_{r,C_m} (cyan circle). The pseudo-plateau bursting trajectory (PPB) is superimposed and has two rotations. From Teka et al. (2011a)

The singular strong canard perturbs to a *primary strong canard* that moves from S^+_{a,C_m} to S_{r,C_m} with only one twist, or one half rotation. In addition, there is a family of *secondary canards* that move through the funnel and exhibit rotations as they flow from S^+_{a,C_m} to S_{r,C_m}. The maximum number of rotations produced, S_{max}, is determined by the eigenvalue ratio of the linearization at the folded node. If λ_s and λ_w are the strong and weak eigenvalues of the linearization at the FN, then define

$$\mu = \frac{\lambda_w}{\lambda_s}.$$ (1.37)

The maximum number of oscillations is then (Rubin and Wechselberger (2008), Wechselberger (2005))

$$S_{max} = \left[\frac{\mu + 1}{2\mu} \right]$$ (1.38)

which is the greatest integer less than or equal to $\frac{\mu+1}{2\mu}$. For $C_m > 0$, but small, there are $S_{max} - 1$ secondary canards that divide the funnel into S_{max} sectors (Brøns et al. (2006)). The first sector is bounded by SC and the first secondary canard ξ_1 and trajectories entering this sector have one rotation. The second sector is bounded by ξ_1 and ξ_2 and trajectories entering here have two rotations, etc. Trajectories entering the last sector, bounded by the last secondary canard and the fold curve L^+, have the maximal S_{max} number of rotations (Rubin and Wechselberger (2008), Vo et al. (2010), Wechselberger (2005)). Many of these small oscillations are so small that they would be practically invisible, particularly in an experimental voltage trace where they would be obscured by noisy fluctuations.

Figure 7 shows a portion of the pseudo-plateau burst trajectory (PPB, black curve) superimposed onto the twisted slow manifold. Since it enters the funnel between the first and second secondary canards it exhibits two rotations as it moves through the region near the FN. These rotations are the small spikes that occur during the active phase of the burst. The full burst trajectory, then, consists of slow flow along the lower and upper sheets of the slow manifold, followed by fast jumps from one attracting sheet to another. The jump from S^+_{a,C_m} down to S^-_{a,C_m} is preceded by a few small oscillations, which are the spikes of the burst. As C_m is made smaller, the burst trajectory looks more and more like the singular periodic orbit, and indeed the small oscillations disappear in the singular limit (Vo et al. (2010)).

4.3 Phase-Plane Analysis of the Desingularized System

Because the desingularized system is two-dimensional, one can apply phase-plane analysis techniques to it (Rubin and Wechselberger (2007), Teka et al. (2011a)). This is illustrated in Fig. 8, where the nullclines and equilibria are shown. The V-nullcline satisfies $F(V, c) = 0$ and is the single z-shaped curve in the figure. The c-nullcline satisfies

$$f_c(\alpha I_{Ca} + k_c c)\frac{\partial f}{\partial V} = 0 \qquad (1.39)$$

and thus

$$\alpha I_{Ca} + k_c c = 0 \qquad (1.40)$$

or

$$\frac{\partial f}{\partial V} = 0. \qquad (1.41)$$

The first set of solutions forms the c-nullcline of the full system and is labelled CN1 in Fig. 8. The second set of solutions forms the two fold curves L^+ and L^-. Intersections of the V-nullcline with CN1 produce ordinary singularities and are equilibria of the full system (Eqs. 1.1–1.3). There is one such equilibrium in Fig. 8A, labelled as point A, which is an unstable saddle point of the desingularized system. Intersections of the V-nullcline with one of the fold curves produce folded

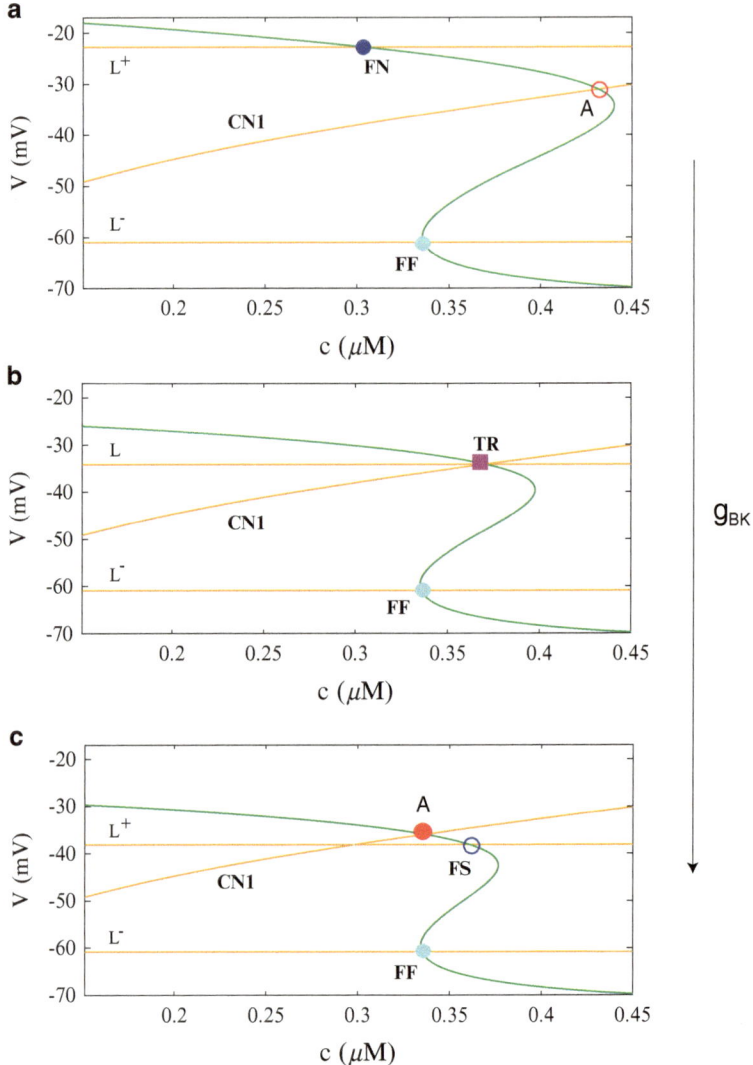

Fig. 8 Nullclines of the desingularized system. (**A**) An ordinary singularity (point A) occurs where the V-nullcline and the CN1 branch of the c-nullcline intersect. This equilibrium is a saddle point of the desingularized system and a saddle-focus of the full system. Two folded equilibria occur where the V-nullcline intersects the fold curves. One folded singularity is a stable folded node (FN), while the other is a stable folded focus (FF). (**B**) When g_{BK} is increased from 0.4 nS to 2.176 nS the saddle point and folded node coalesce at a transcritical bifurcation (TR). This is also known as a folded saddle-node of type II. (**C**) When $g_{BK} = 4$ nS the ordinary singularity, which now occurs on the top sheet of the critical manifold, is stable. The folded node has become a folded saddle and is unstable

singularities. In Fig. 8A there is a folded focus singularity on L^- and a folded node singularity on L^+. The folded node is stable, and will generate canards. The folded focus is also stable, but it produces no canards.

One advantage of having a planar system is that it facilitates understanding of the effects of parameter changes. For example, increasing the parameter g_{BK} changes the shape of the V-nullcline and brings L^+ and L^- closer together, but has no effect on CN1. As this parameter is increased the FN and the equilibrium point A move closer together, and eventually coalesce (Fig. 8B). When the parameter is increased further the stability is transferred from the folded node to the full-system equilibrium (Fig. 8C). Thus, the desingularized system undergoes a transcritical bifurcation as g_{BK} is increased. On the other side of the bifurcation, the folded node has become a folded saddle and no longer attracts trajectories off of its one-dimensional stable manifold. The intersection point A is now stable, and is a stable equilibrium of the full system of equations. Thus, beyond the transcritical bifurcation the full system is at rest at a high-voltage (depolarized) steady state. This transcritical bifurcation of the desingularized system is also called a *type II folded saddle-node bifurcation* (Krupa and Wechselberger (2010), Milik and Szmolyan (2001), Szmolyan and Wechselberger (2001)). In contrast, a *type I folded saddle-node bifurcation* is the coalescence of a folded saddle and a folded node singularity, and does not involve full-system equilibria (Szmolyan and Wechselberger (2001)).

The transcritical bifurcation of the desingularized system is a signature of a *singular Hopf bifurcation* of the full system (Desroches et al. (2012), Guckenheimer (2008)). The ordinary saddle point of the desingularized system in Fig. 8A is a saddle focus of the full system, and trajectories can approach the saddle focus along its one-dimensional stable manifold and leave along the two-dimensional unstable manifold with growing oscillations. In fact, with an appropriate global return mechanism, this can be a mechanism for MMOs that is different from that due to the folded node (which co-exists with the saddle focus). In this case, the small oscillations are characterized by a monotonic increasing amplitude, which may or may not be the case for canard-induced MMOs. Interestingly, these two mechanisms for MMOs are not mutually exclusive; in Fig. 21 of Desroches et al. (2012) an example is shown of an MMO whose first few small oscillations are due to a twisted slow manifold induced by a folded node and whose remaining small oscillations are due to growing oscillations away from a saddle focus.

4.4 Bursting Boundaries

One useful application of the 1-fast/2-slow analysis is the determination of the region of parameter space for which bursting occurs. A change in a parameter can convert bursting to spiking, as in Fig. 5, or can convert bursting to a stable steady state, as would occur in Fig. 8. Since the pseudo-plateau bursting is closely associated with the existence of a folded node singularity, one necessary condition for this type of bursting is the existence of a folded node. We have seen that a folded node can be created/destroyed via a type II folded saddle-node bifurcation. That is, when the weak eigenvalue crosses through the origin, and thus $\mu = 0$. A folded node can also change to a folded focus, which has no canard solutions. This occurs after the eigenvalues coalesce, i.e., when $\mu = 1$. Since a folded node singularity exists only when $0 < \mu < 1$, canard-induced mixed-mode oscillations only occur for parameter values for which $0 < \mu < 1$ at the folded singularity. This is predictive for pseudo-plateau bursting, at least in the case where C_m is small. For larger values of C_m the singular theory may not hold up, so bursting may occur for parameter values at which the singular theory predicts a continuous spiking solution.

Another condition for canard-induced MMOs is that there is a global return mechanism that periodically injects the trajectory into the funnel. When this occurs, the trajectory moves through the twisted slow manifold and produces small oscillations that are the spikes of pseudo-plateau bursting. If instead the trajectory is injected outside of the funnel, on the other side of the strong canard, continuous spiking will occur. To quantify this, a distance measure δ is used. This is defined using the singular periodic orbit, and is best viewed in the c-V plane (Fig. 9). When the orbit jumps from the bottom sheet of the critical manifold at L^- it moves along a fast fiber to a point on $P(L^-)$ on the top sheet. The horizontal distance from this point to the strong canard is defined as δ. If the point is on the strong canard, then

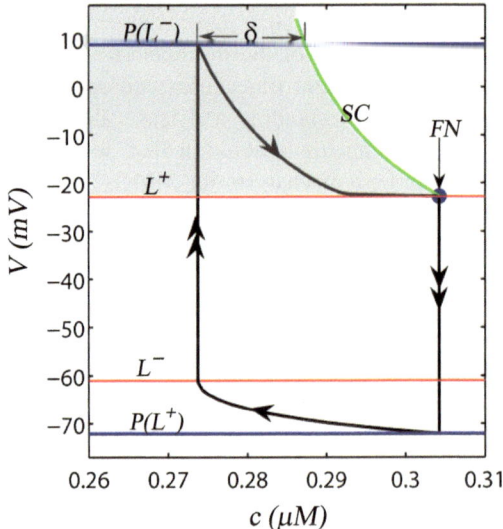

Fig. 9 Projection of the singular periodic orbit and key structures onto the c-V plane. The upper fold curve (L^+) and strong canard (SC) delimit the singular funnel. The singular periodic orbit jumps from L^- onto a point on $P(L^-)$. The distance in the c direction from this point to the strong canard is defined as δ, and by convention $\delta > 0$ when the point is in the funnel. From Teka et al. (2011a)

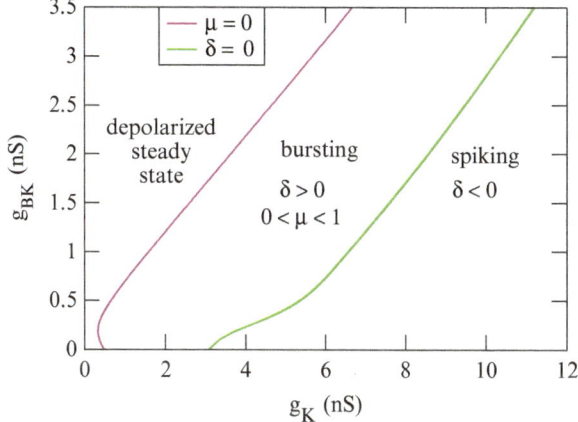

Fig. 10 The singular analysis predicts whether the full system should be continuously spiking, bursting, or in a depolarized steady state. The folded node becomes a folded saddle above the $\mu = 0$ curve and the ordinary singularity of the desingularized system becomes stable. Between the $\mu = 0$ and $\delta = 0$ curves the two conditions are met for mixed-mode oscillations, and pseudo-plateau bursting is predicted to occur. Below the $\delta = 0$ curve the singular periodic orbit does not enter the singular funnel, resulting in relaxation oscillations. Away from the singular limit (for $C_m > 0$) these become a periodic spike train

$\delta = 0$, while if it is in the funnel then $\delta > 0$ by convention. Thus, a necessary condition for the existence of canard-induced MMOs, and pseudo-plateau bursting, is $\delta > 0$.

With these constraints on δ and μ one can construct a 2-parameter bifurcation diagram characterizing the behavior of the full system. One such diagram is illustrated in Fig. 10, where the maximum conductances of the delayed rectifier (g_K) and the large-conductance K(Ca) (g_{BK}) currents are varied. In the diagram, the upper curve (magenta) consists of type II folded saddle-node bifurcations that give rise to a folded node, and thus is characterized by $\mu = 0$. Above this curve the full system equilibrium is stable and the system goes to a depolarized steady state. Below this curve $\mu > 0$. The lower curve (green) consists of points in which $\delta = 0$. Above this curve $\delta > 0$, while below it $\delta < 0$. Both conditions for MMOs are satisfied between the two curves, so this is the parameter region where mixed-mode oscillations occur.

4.5 Spike-Adding Transitions

In the region of parameter space where MMOs occur, one can characterize the number of small oscillations (spikes) that occur in different subregions. Such an analysis was performed in Vo et al. (2012), using a variant of the lactotroph model (described in the Appendix) that we have been using thus far. It was motivated by the observation that, in a 4-variable lactotroph model containing an A-type K$^+$ current, pseudo-plateau bursting can occur even if one fixes the c variable at its average value (Toporikova et al. (2008)). Thus, to simplify the analysis, c is clamped and the model reduced to 3 dimensions. This 3-dimensional model is what we consider now, where the major difference with the 3-dimensional lactotroph model discussed previously is that the SK and BK currents are replaced by leakage and A-type K^+ currents, and the calcium variable c is replaced by an inactivation variable e for the A-type channels. The bursting boundaries were determined with this model in the plane of the two parameters g_K and g_A. In this case, the left bursting boundary occurs when $\mu = 0$ and the folded node becomes a folded saddle at a type II folded saddle-node bifurcation. Unlike in Fig. 10, however, the right boundary for mixed-mode oscillations occurs when $\mu = 1$ and the folded node becomes a folded focus (Fig. 11). A third boundary occurs where $\delta = 0$, and the fourth boundary occurs where a stable equilibrium of the full system is born at a saddle-node on invariant circle (SNIC) bifurcation. Both conditions for MMOs are satisfied within the trapezoidal region bounded by these line segments (Fig. 11).

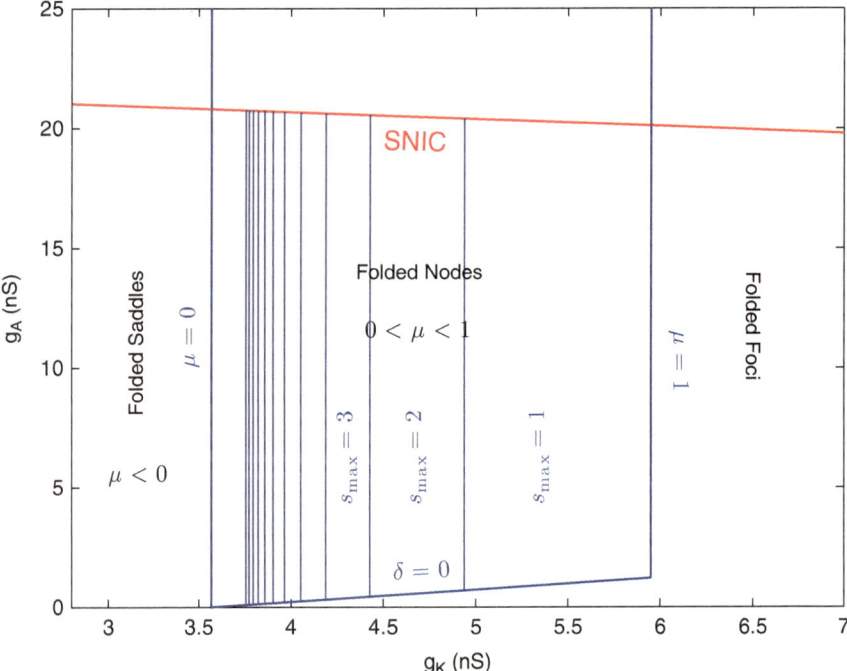

Fig. 11 Bursting boundaries and the maximum number of spikes per burst in a variant of the lactotroph model (described in Appendix). The left and right boundaries occur when the folded node becomes a folded saddle ($\mu = 0$) or a folded focus ($\mu = 1$). The lower boundary occurs when the periodic orbit jumps to the strong canard that delimits the singular funnel ($\delta = 0$). The upper boundary occurs when a stable equilibrium of the full system is born at an SNIC bifurcation. The maximum number of spikes (S_{max}) is determined by μ. From Vo et al. (2012)

The maximum number of small oscillations that occur in the mixed-mode oscillations (S_{max}) depends on μ, the eigenvalue ratio, according to Eq. 1.38. In this model, the eigenvalues depend only on g_K, and only slightly on g_A. Thus, the subregions of constant S_{max} are separated by almost-vertical line segments (g_K values where the value of the greatest integer function changes). Near the right boundary $\mu \approx 1$, so by Eq. 1.38 there is at most one small oscillation per burst. (There will be an additional oscillation, due to the trajectory jumping from the lower sheet to the upper sheet of the slow manifold; after the jump, the voltage is initially large and then slowly declines, producing the first spike of the burst.) For $g_K \approx 5$ nS, S_{max} increases to 2, and then to 3 for $g_K \approx 4.4$ nS. The maximum number of oscillations continues to increase as the left boundary is approached, where $\mu = 0$ and $S_{max} \to \infty$.

While the eigenvalue ratio tells half of the story, the other half is determined by where the periodic orbit lands when it jumps to the top sheet of the slow manifold (i.e., it depends on the value of δ). If the orbit jumps to a point close to the primary strong canard, then δ is near 0. In this sector, bounded on one side by the primary strong canard and on the other by the first secondary canard, one small oscillation will be produced, regardless of the eigenvalue ratio μ. This is the case near the bottom of the MMO region in the parameter plane. The distance measure δ becomes larger for larger values of g_A, and thus the number of small oscillations produced during a burst increases as the trajectory jumps into sectors that are further from the primary strong canard. In summary, the parameter g_K controls the eigenvalue ratio μ and thus the maximum number of spikes per burst. It also determines the number of secondary canards, which delimit sectors of the funnel. The parameter g_A controls the distance measure δ and thus which sector the orbit jumps into when it jumps to the top sheet of the slow manifold. In the two-parameter diagram of Fig. 11, the number of spikes per burst will increase as one moves to the left or upward in the MMO region.

If δ is held constant by fixing g_A, and μ is varied by varying g_K, what will the bifurcation structure of the spike adding transitions look like? How are the MMO solution branches connected to one another? That is, how does a bursting branch with n spikes connect to a bursting branch with $n + 1$ spikes? These questions were addressed in Vo et al. (2012), first by performing a bifurcation analysis with the continuation program AUTO (Doedel (1981), Doedel et al. (2007)), and then by using return maps of both the singular and non-singular systems to better understand the spike adding behavior. Figure 12 shows the L_2 norm of the solution over a range of values of g_K for $g_A = 4$ nS. For g_K below about 3.7 nS there is a stable depolarized steady state (E_D). This becomes unstable at a subcritical Hopf bifurcation. The family of periodic solutions born at this bifurcation consists of continuous spiking, labeled here as $s = 0$ (no small oscillations). The first family of bursting solutions ($s = 1$ branch) connects to the spiking branch at a period doubling bifurcation (at $g_K \approx 3.592$ nS, shown in the left inset) and later at a second period doubling bifurcation (at $g_K \approx 6.127$ nS). This bursting pattern has one spike induced by a folded node, in addition to the initial spike due to the jump up to and initial motion down the top sheet. The next bursting branch, with $s = 2$, is

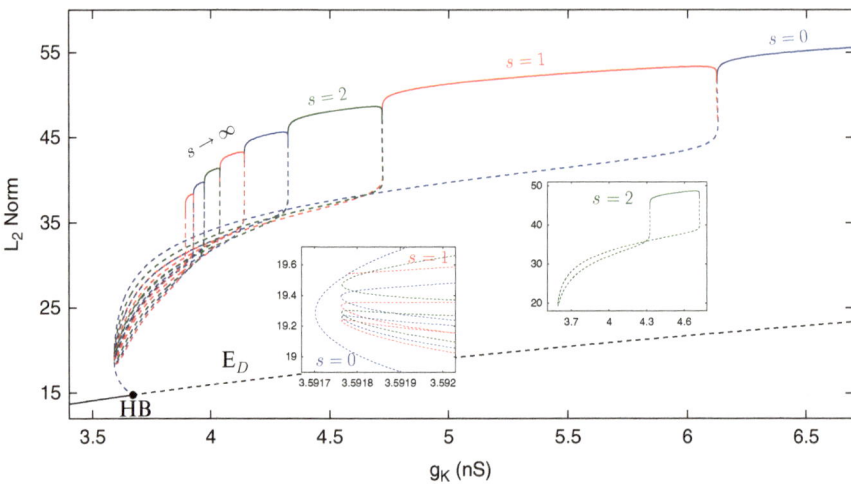

Fig. 12 Spike adding transitions over a range of values of g_K and with $g_A = 4$ nS and $C_m = 2$ pF. Only the $s = 1$ bursting family is connected to the spiking branch ($s = 0$). All other families of bursting solutions (e.g., the $s = 2$ family) form isolas. The lactotroph model variant is used. From Vo et al. (2012)

connected to neither the spiking branch nor the $s = 1$ branch. Instead, it is an isola formed by a pair of saddle-node of periodics bifurcations (right inset of Fig. 12). This family of solutions extends over a smaller range of g_K values than the $s = 1$ family, and the stable portion of the branch in particular is only about a quarter as long as that of the $s = 1$ branch. Other bursting families are isolas similar to the $s = 2$ family, and the range of each successive family is shorter than its predecessor. There is an accumulation point as g_K is decreased (toward the point where $\mu = 0$) as the stable range of the bursting families approaches 0 and $s \to \infty$.

4.6 *Prediction Testing on Real Cells*

Figures 10 and 11 provide predictions about how the number of spikes per burst vary with parameter values and the boundaries between continuous spiking, bursting, and stationary behavior. These predictions have proven to be quite good (Teka et al. (2011a), Vo et al. (2010)) even in cases where the singular parameter C_m is large, within the right range for pituitary cells (≈ 5 pF). Importantly, these predictions also apply to real pituitary cells. For example, if a pituitary cell is spiking continuously, then it should be possible to convert it to a bursting cell by increasing the conductance of the BK-type or A-type K$^+$ currents, or by decreasing the conductance of the delayed-rectifier K$^+$ current. Also, if the cell is bursting, then the number of spikes in a burst should increase if g_{BK} or g_A is increased, or if g_K is decreased. These predictions were tested using the dynamic clamp technique, which records the voltage from a real cell using an electrode, then uses a mathematical model to compute a current, which is injected into the cell in real time. Thus, the dynamic clamp allows one to add a voltage-dependent current to a real cell, using the cell's membrane potential to calculate that current (Milescu et al. (2008), Sharp et al. (1993)).

Figure 13 shows the result of adding a BK current to a GH4C1 cell (a lacto-somatotroph cell line). With no added BK conductance the cell spikes continuously

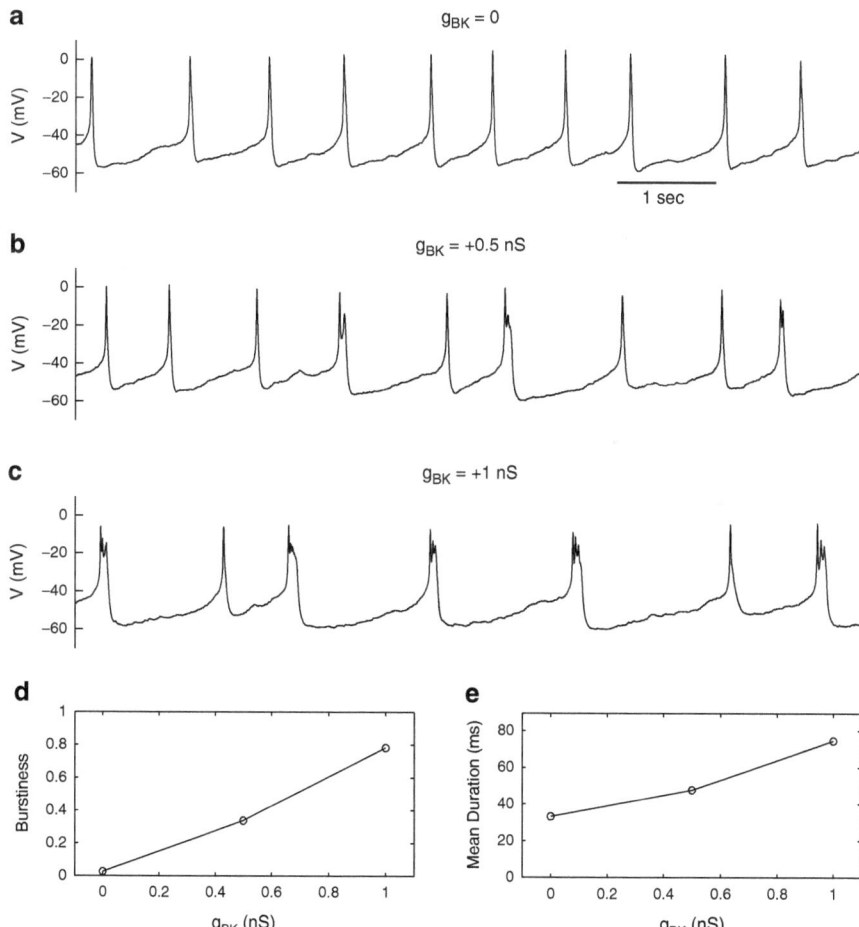

Fig. 13 Patch clamp recording from a GH4C1 lacto-somatotroph cell using dynamic clamp to add a model BK-type current (Eq. 1.10). (**A**) With no added current the cell spikes continuously. (**B**) When 0.5 nS of BK conductance is added the cell exhibits bursts intermingled with spikes. (**C**) With a larger added BK conductance, 1 nS, the burstiness is increased, as is the number of spikes per burst. (**D**) Quantification of the burstiness over the entire time course for the three values of the added g_{BK}. The burstiness increases with g_{BK}. (**E**) Quantification of the mean event duration for the three values of the added g_{BK}. The event duration increases with g_{BK}

(Fig. 13A). However, once BK current is added with a sufficiently large conductance the cell exhibits a pseudo-plateau bursting pattern intermixed with spiking (Fig. 13B). Adding more conductance increases the burstiness of the cell, that is, the fraction of events that are bursts. Also, as predicted by the analysis, adding more g_{BK} increases the number of spikes in a burst (Fig. 13C). The change in burstiness with added BK conductance is quantified in panel D, where the burstiness is calculated over the entire time course for each value of the added g_{BK}. Panel E shows the quantification of the mean even duration, including both spikes and bursts. Both the burstiness and the event duration increase with an increase in the added g_{BK}, as predicted by the analysis. The transition between spiking and bursting with addition or subtraction of a BK current using the dynamic clamp was shown repeatedly in GH4C1 cells and primary pituitary gonadotrophs (Tabak et al. (2011), Tomaiuolo et al. (2012)).

It is also possible to use the dynamic clamp to add a negative conductance to the cell, thereby subtracting an ionic current. One can, for example, develop a model for the I_K current that reflects the characteristics of this current in the real cell. Then the dynamic clamp technique can be used to subtract off some of this current from the cell, by adding a negative g_K conductance. This can be superior to using pharmacological agents to remove a current, since such agents are often non-specific. Also, the dynamic clamp approach allows the investigator to subtract off only a fraction of the current, in a controlled manner. We use this approach to subtract off g_K conductance, and thus reduce the effective g_K value in the cell (the native g_K minus that subtracted off with dynamic clamp). The prediction is that a spiking cell should become a burster when a sufficient amount of g_K is subtracted, and as more is subtracted the number of spikes in a burst should increase (Fig. 12). The results of applying the dynamic clamp to a GH4C1 cell are shown in Fig. 14. The top panel shows that the cell is mostly spiking, with a low degree of burstiness and no more than 2 spikes per burst, prior to subtraction of g_K. When some delayed rectifier conductance is subtracted (-1 nS) the burstiness of the cell increases (Fig. 14B). Subtracting off even more conductance (-2 nS) further increases the burstiness and increases the number of spikes per burst (Fig. 14C). These effects are quantified in panels D and E, where burstiness and mean even duration are computed over the entire time course durations. As predicted by the singular analysis, reducing g_K in the cell increases the likelyhood that it wil burst, and increases the number of spikes within a burst.

5   Relationship Between the Fast/Slow Analysis Structures

We began with a description of the standard fast/slow analysis technique applied to bursting oscillations in which the full 3-dimensional system is decomposed into a 2-dimensional fast subsystem and a 1-dimensional slow subsystem. We then described an alternate decomposition, with a 1-dimensional fast subsystem and a 2-dimensional slow subsystem. Each approach made use of key structures that

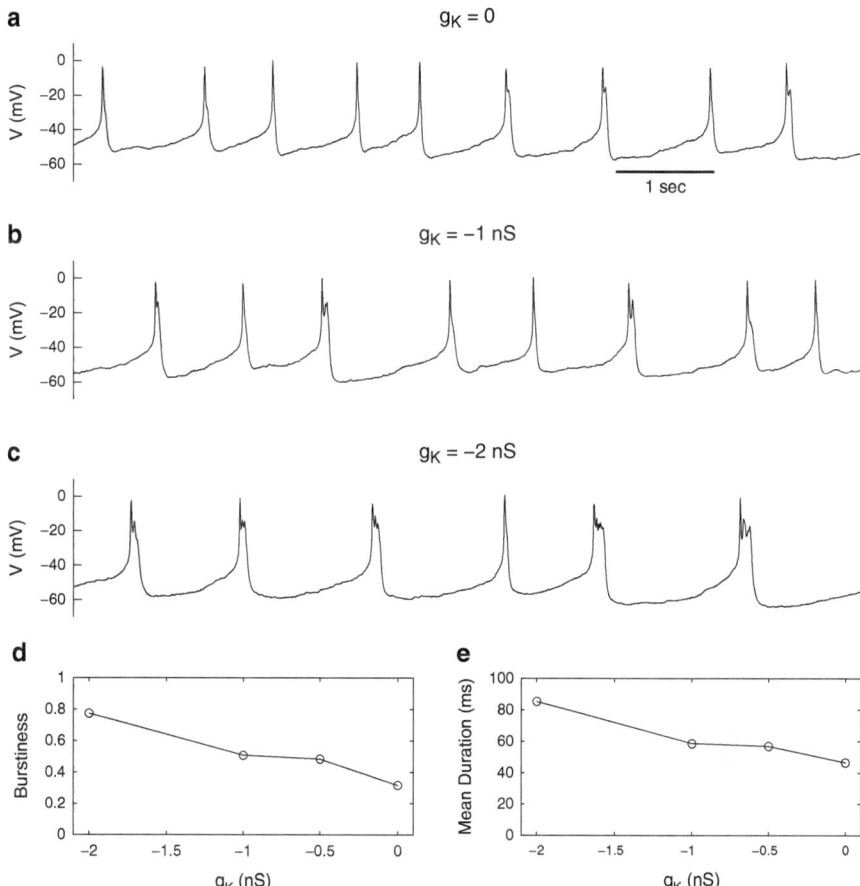

Fig. 14 Patch clamp recording from a GH4C1 lacto-somatotroph cell using dynamic clamp to subtract a delayed rectifier K$^+$ current (Eq. 1.6). (**A**) With no added current the cell has very low burstiness, with at most two spikes per burst. (**B**) When 1 nS of delayed rectifier K conductance is subtracted the burstiness of the cell increases. (**C**) When more delayed rectifier conductance is subtracted, 2 nS, the burstiness increases further and the number of spikes per burst increases. (**D**) Quantification of the burstiness over the entire time course for the three values of the subtracted g_K. The burstiness increases when more g_K is subtracted. (**E**) Quantification of the mean event duration for the three values of the subtracted g_K. The event duration increases when more g_K is subtracted

organized the behavior of the system. In the case of the 2-fast/1-slow analysis (Fig. 3), the z-curve (equilibria of the fast subsystem) and the subcritical Hopf bifurcation point on the upper portion of the z-curve are two key structures. In addition, the nullcline of the slow variable (the c-nullcline) is important since it determines the direction of the slow flow. In the case of the 1-fast/2-slow analysis, the critical manifold, the folded node singularity, and the nullclines of the desingularized system are key organizational structures (Figs. 6, 8). In Teka et al.

(2012) the lactotroph model (with SK and BK currents, and calcium variable c) was used to investigate the relationship between these sets of structures. In this section we discuss the key findings of this investigation.

5.1   The $f_c \to 0$ Limit

The nullclines of the desingularized system shown in Fig. 8A are redrawn in Fig. 15A. These were computed using $f_c = 0.01$, which is the typical value for this parameter (the ratio of free to bound Ca^{2+} in the cell). Superimposed is the z-curve obtained from the 2-fast/1-slow decomposition, computed using $C_m = 10$ pF. This z-curve is the stationary branch of the 2-variable fast subsystem, where c is treated as a parameter, so it tacitly assumes that $f_c = 0$. It is clear that the V-nullcline and z-curve are very similar, and the CN1 nullcline of the desingularized system is the c-nullcline of the 2-fast/1-slow system. The point A is the single intersection of all three curves, and is both an equilibrium of the desingularized system and an equilibrium of the full system. The subcritical Hopf bifurcation of the 2-fast/1-slow system lies on the top branch of the z-curve, but below L^+, which means that it is located on the middle sheet of the critical manifold (Fig. 15B). In addition, the two saddle-node bifurcations of the z-curve are on the middle sheet of the critical manifold. The folded node of the desingularized system is located close to the subcritical Hopf bifurcation point, but on the upper fold of the critical manifold.

We now take the limit $f_c \to 0$, so that the variable c becomes infinitesimally slow. Taking this limit has no effect on the z-curve, which already assumes that $f_c = 0$. It also has no effect on L^+, L^-, or CN1, since f_c divides out of the equations for these curves. However, it does influence the V-nullcline of the desingularized system,

$$-f_c(\alpha I_{Ca} + k_c c)\frac{\partial f}{\partial c} + \left(\frac{n_\infty(V) - n}{\tau_n}\right)\frac{\partial f}{\partial n} = 0. \qquad (1.42)$$

When $f_c \to 0$ the first term disappears, and for the second term to equal 0 either $n = n_\infty(V)$ or $\frac{\partial f}{\partial n} = g_k(V - V_K) = 0$. Since $V > V_K$, and $g_K \neq 0$, we must have $n = n_\infty(V)$, so that $\frac{dn}{dt} = 0$. Thus, the V-nullcline of the desingularized system satisfies $\frac{dV}{dt} = 0$ and $\frac{dn}{dt} = 0$, which are the same equations defining the z-curve.

Although the V-nullcline and the z-curve superimpose in the $f_c \to 0$ limit, the folded node and the Hopf bifurcation do not (Fig. 16). Instead, in this limit, the Hopf bifurcation remains on the middle sheet of the critical manifold.

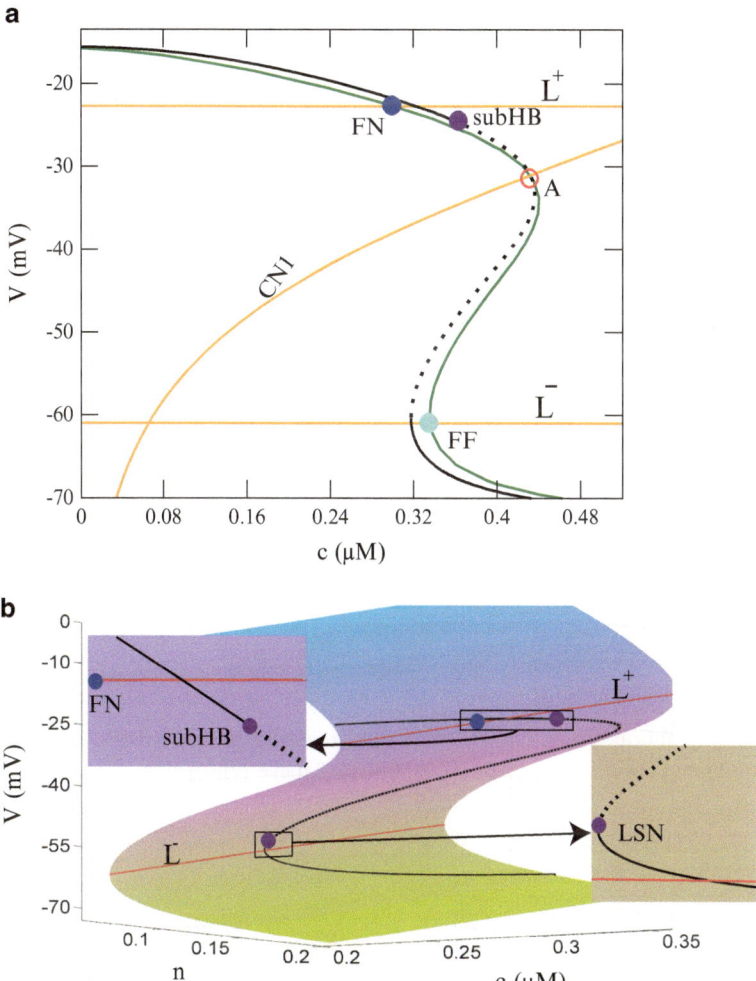

Fig. 15 (**A**) The nullclines of the desingularized system with the z-curve (black) superimposed. (**B**) The critical manifold of the reduced system with the z-curve superimposed. In both cases, $g_K = 4$ nS, $g_{BK} = 0.4$ nS, $C_m = 10$ pF (for the z-curve), and $f_c = 0.01$ (for the desingularized system). Redrawn from Teka et al. (2012)

5.2 The $C_m \to 0$ Limit

While the limit $f_c \to 0$ makes c infinitesimally slow, the limit $C_m \to 0$ makes V infinitely fast. We now take this limit, returning f_c to its default value of 0.01. The desingularized system is formed from the limit $C_m \to 0$, so taking this limit only affects the z-curve of the 2-fast/1-slow decomposition. This curve of fast-subsystem equilibria is defined by $f(V, n, c) = 0$ and $n = n_\infty(V)$, and C_m appears

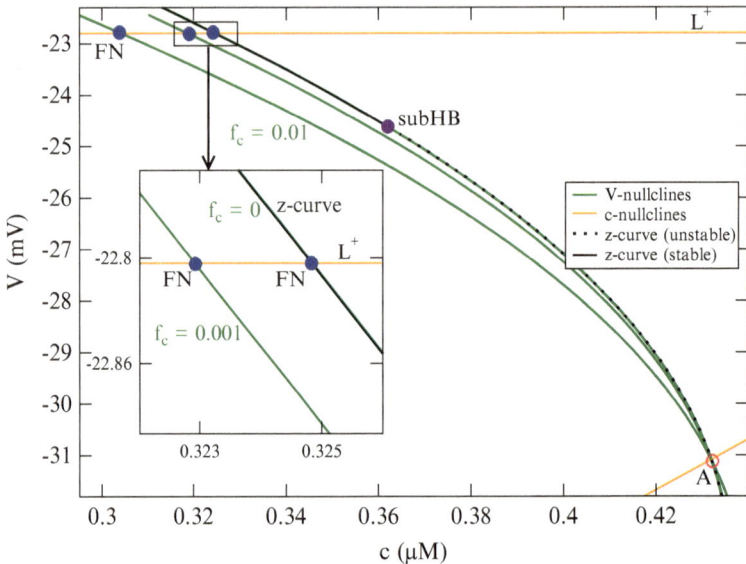

Fig. 16 The V-nullcline (green) converges to the z-curve (black) in the $f_c \to 0$ limit. The folded node and subcritical Hopf bifurcation remain separated. The z-curve was computed using $C_m = 10$ pF. From (Teka et al. (2012))

in neither equation. Thus, the locations of the equilibria that comprise the z-curve are unaffected by C_m. However, the stability of these points does change with C_m, since C_m is in the ordinary differential equation for V (Eq. 1.1). In fact, as $C_m \to 0$ the Hopf bifurcation migrates toward the fold curve L^+ (Fig. 17).

To understand this convergence to L^+, note that the Jacobian matrix of the 2-dimensional fast subsystem (Eqs. 1.1, 1.2) is

$$\mathbf{J} = \begin{pmatrix} \frac{1}{C_m}\frac{\partial f}{\partial V} & \frac{1}{C_m}\frac{\partial f}{\partial n} \\ \frac{\partial g}{\partial V} & \frac{\partial g}{\partial n} \end{pmatrix} \tag{1.43}$$

where $g(V) \equiv \frac{n_\infty(V)-n}{\tau_n}$. The trace of $\mathbf{J}$ is

$$\text{trace}(\mathbf{J}) = \frac{1}{C_m}\frac{\partial f}{\partial V} + \frac{\partial g}{\partial n} \tag{1.44}$$

and at a Hopf bifurcation trace$(\mathbf{J}) = 0$. Thus, at the Hopf,

$$\frac{\partial f}{\partial V} + C_m\frac{\partial g}{\partial n} = 0 \ . \tag{1.45}$$

In the $C_m \to 0$ limit the second term disappears, requiring that $\frac{\partial f}{\partial V} = 0$. This is the equation for the fold curve.

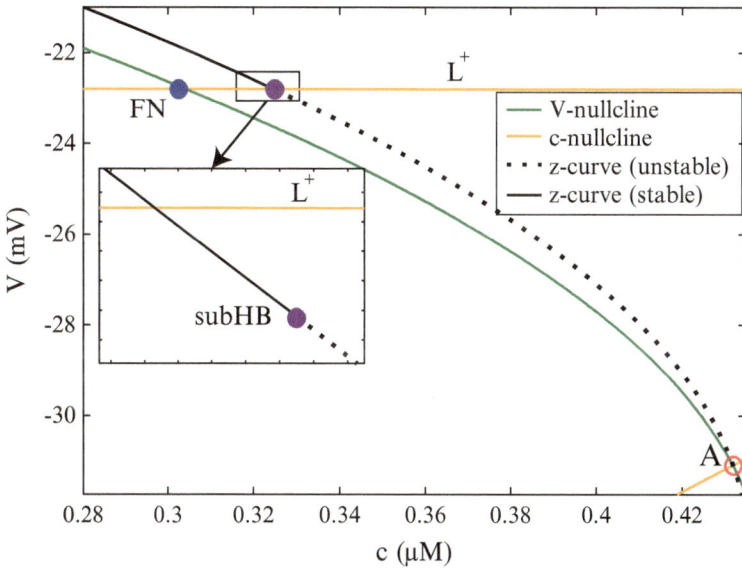

Fig. 17 In the $C_m \to 0$ limit the Hopf bifurcation on the z-curve migrates to the fold curve L^+. In this figure $C_m = 0.1$ pF, so it is very close to L^+, but has not yet reached it (inset). The V-nullcline of the desingularized system is computed with the default $f_c = 0.01$. From Teka et al. (2012)

5.3 The Double Limit

In the $C_m \to 0$ limit the Hopf bifurcation point migrated to the upper fold curve, but remained distinct from the folded node singularity since the V-nullcline of the desingularized system does not overlay the z-curve. The two coalesce when, in addition to taking $C_m \to 0$, one takes the $f_c \to 0$ limit. In this double limit the V-nullcline converges to the z-curve and the Hopf bifurcation is on the fold curve L^+, and thus the folded node singularity of the desingularized system and the Hopf bifurcation of the 2-dimensional fast subsystem of the 2-fast/1-slow decomposition are the same point.

It is interesting to see how the bursting orbit changes as the double limit is approached from the f_c direction and from the C_m direction. Fig. 18A shows the bursting orbit computed with $f_c = 0.01$ and $C_m = 10$ (within the range of values for a pituitary lactotroph or somatotroph), superimposed with the V-nullcline of the desingularized system and the z-curve. In this case, the system is far from any singular limit, so the orbit is only somewhat close to the z-curve and the spikes are large. When f_c is reduced by a factor of 10 the bursting orbit (which is actually more like a relaxation oscillation) is clearly organized by the z-curve (Fig. 18B). During the silent phase it moves along the bottom branch, while during the active phase it moves along the top branch. It passes through the subcritical Hopf bifurcation, and follows the unstable branch for some time before moving away with oscillations of

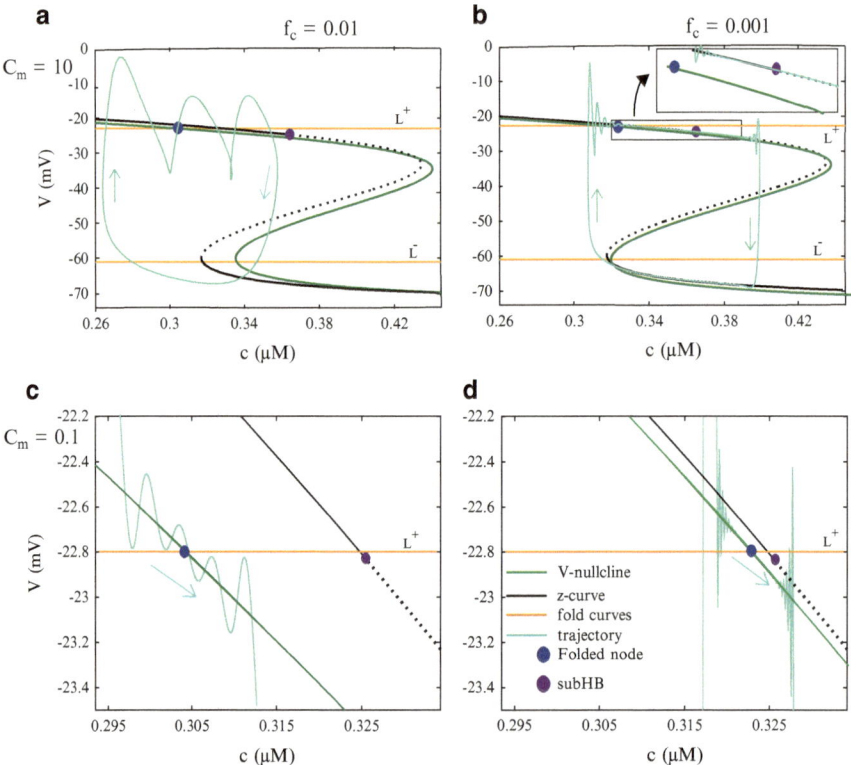

Fig. 18 Bursting orbits together with the z-curve and V-nullcline of the desingularized system for different approaches to the double limit. (**A**) With physiological values of f_c and C_m the orbit (cyan) follows neither the z-curve nor the V-nullclline and the spikes are relatively large. (**B**) As the double limit is approached in the f_c direction the bursting orbit follows the z-curve. (**C**) As the double limit is approached in the C_m direction the orbit follows the V-nullcline and passes through the folded node. (**D**) Near the double limit the orbit passes through the folded node, but also moves close to the z-curve and the Hopf bifurcation. From Teka et al. (2012)

increasing size. Thus, it exhibits the slow passage effect that is well documented for an orbit of a fast/slow system as it moves through a subcritical Hopf bifurcation (Baer et al. (1989), Baer and Gaekel (2008)). If f_c is returned to its original value and C_m is reduced by a factor of 100, then the system is organized by the structures of the desingularized system. Fig. 18C shows that in this case the burst trajectory passes very close to the folded node singularity as it moves along the V-nullcline. The spikes are small, and first decrease and then increase in amplitude as the orbit moves along the twisted slow manifold, which is typical for passage near a folded node singularity (Desroches et al. (2012)). If C_m is kept at this small value and f_c is now reduced by a factor of 10, then the bursting orbit again moves through the folded node along the V-nullcline, but this time with more spikes and a much more extreme decrease in amplitude near the folded node. Also, since this is near the double limit, the trajectory passes near the z-curve, and the folded node and Hopf bifurcation are close together.

6 Store-Generated Bursting in Stimulated Gonadotrophs

The pseudo-plateau bursting that we have discussed so far is common in the spontaneous activity of pituitary somatotrophs and lactotrophs, and is sometimes observed in the spontaneous activity of gonadotrophs (Stojilković et al. (2010)). More often, though, gonadotrophs exhibit a tonic spiking pattern that yields little hormone release (Van Goor et al. (2001b)). However, when stimulated by the physiological stimulator *gonadotropin-releasing hormone* (GnRH) the gonadotrophs typically produce a bursting pattern with period of roughly 4–15 sec that results in a much higher level of luteinizing hormone release (Stojilković et al. (2010)). This was first observed using Ca^{2+} imaging, where a train of Ca^{2+} spikes was observed in the presence of GnRH (Shangold et al. (1988)). In a series of papers published in the 1990s, it was shown that this bursting pattern is due to the interaction of a Ca^{2+} oscillator stimulated by GnRH and an electrical oscillator that produces tonic spiking when the Ca^{2+} oscillator is turned off (Kukuljan et al. (1994), Stojilković et al. (1992, 1993), Stojilković and Tomić (1996), Tse and Hille (1992), Tse et al. (1994, 1997)). A key element of this research was the development of a mathematical model that helped with the interpretation of the data and guided experiments (Li et al. (1994, 1995), Rinzel et al. (1996)). In this section we discuss a simplified version of this model that retains the key biophysical and dynamic elements (Sherman et al. (2002)).

6.1 Closed-Cell Dynamics

We begin with the Ca^{2+} oscillator. Here, oscillations in the free cytosolic Ca^{2+} concentration (c) are due to the cycling of Ca^{2+} into and out of the endoplasmic reticulum (ER). Denote the ER Ca^{2+} concentration as c_{ER}. In the closed-cell model we analyze first, the total number of Ca^{2+} ions in the cell is conserved; ions simply move back and forth between the cytosolic compartment and the ER compartment. Denote the total free Ca^{2+} concentration in the cell as c_{tot}. Then

$$c_{tot} = c + \sigma c_{ER} \tag{1.46}$$

where σ is the ratio of the "effective ER volume" $\bar{V}_{ER} = \frac{V_{ER}}{f_{ER}}$ (V_{ER} is the ER volume and f_{ER} is the fraction of unbound Ca^{2+} in the ER), to the "effective cytosol volume" $\bar{V}_c = \frac{V_c}{f_c}$. That is,

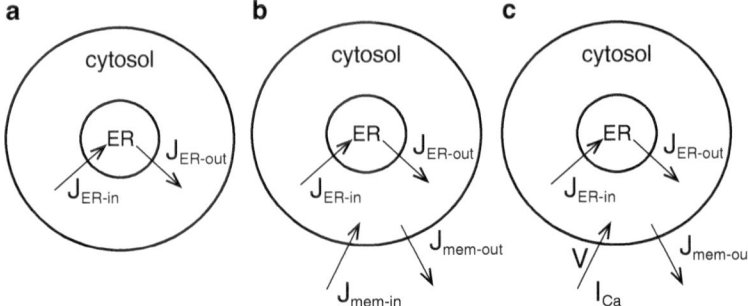

Fig. 19 Diagram of the Ca^{2+} fluxes in (**A**) the closed-cell model, (**B**) the open-cell model with constant influx, and (**C**) the open-cell model with influx through voltage-gated Ca^{2+} channels

$$\sigma = \frac{V_{ER}f_c}{V_c f_{ER}} \quad . \tag{1.47}$$

Rewriting,

$$c_{ER} = (c_{tot} - c)/\sigma \tag{1.48}$$

where c_{tot} is constant. The differential equation for the cytosolic Ca^{2+} concentration is

$$\frac{dc}{dt} = (J_{ER-out} - J_{ER-in})/\bar{V}_c \tag{1.49}$$

where J_{ER-in} and J_{ER-out} are the calcium fluxes into and out of the ER, respectively (Fig. 19A).

The flux of Ca^{2+} into the ER is through pumps powered by the hydrolysis of adenosine triphosphate (ATP). These are called SERCA (Sarcoplasmic-Endoplasmic Reticulum Calcium ATPase) pumps. The pump rate is an increasing function of the cytosolic Ca^{2+} concentration, and in some models includes a dependence on the ER Ca^{2+} concentration (Sneyd et al. (2003)). We use a simple second-order Hill function of c to describe the flux through SERCA pumps,

$$J_{ER-in} = \frac{V_1 c^2}{K_1^2 + c^2} \tag{1.50}$$

where K_1 is the Ca^{2+} concentration for the half-maximal pump rate and V_1 is the maximum pump rate.

The flux of Ca^{2+} out of the ER has two components. First, there is leakage that is assumed to be proportional to the difference in the ER and cytosolic Ca^{2+} concentrations,

$$J_{leak} = L(c_{ER} - c) \ . \tag{1.51}$$

The Ca^{2+} concentration in the ER is greater than that in the cytosol, so the leakage is from the ER into the cytosol. The second component is only active when GnRH binds to receptors in the cell's plasma membrane, generating the intracellular signaling molecule inositol 1,4,5-trisphosphate (IP_3). This molecule binds to IP_3 receptors in the ER membrane and can activate them. Once activated, the IP_3 receptors behave like Ca^{2+} channels, allowing Ca^{2+} to flow out of the ER and into the cytosol down the Ca^{2+} gradient. Cytosolic Ca^{2+} ions can also bind to regulatory sites on the receptor, increasing its open probability. Thus, the IP_3 receptors are gated by both IP_3 and cytosolic Ca^{2+}. There is a third binding site on each receptor subunit, for Ca^{2+}-induced inactivation of the receptor. This negative feedback operates on a slower time scale, so that in the presence of IP_3, Ca^{2+} provides both fast positive feedback and slower negative feedback onto the IP_3 receptor. Thus, the IP_3 receptor has dynamics that are very similar to those of the voltage-gated Na^+ channel that is ubiquitous in neurons, as was demonstrated by Li and Rinzel (1994). In the expression that we use for the probability that the IP_3 receptor/channel is open the kinetics of IP_3 binding and Ca^{2+} binding to the activation site are instantaneous, while the Ca^{2+} binding to the inactivation site occurs with a time constant τ_h. The IP_3 open probability is then multiplied by the Ca^{2+} gradient, which provides the driving force for Ca^{2+} flux:

$$J_{IP3} = P \left(\frac{c^3}{(c + k_a)^3} \right) \left(\frac{IP_3^3}{(IP_3 + k_i)^3} \right) h^3 \, (c_{ER} - c) \tag{1.52}$$

where P is a parameter representing the flux through an open channel, k_a, k_i are parameters, IP_3 is the intracellular IP_3 concentration, and h is an *inactivation variable* satisfying the differential equation

$$\frac{dh}{dt} = (h_\infty - h)/\tau_h \tag{1.53}$$

where

$$h_\infty(c) = \frac{K_d}{K_d + c} \tag{1.54}$$

and

$$\tau_h(c) = \frac{A}{K_d + c} \ . \tag{1.55}$$

Here K_d is the dissociation constant for Ca^{2+} binding to the inactivation site (i.e., the unbinding rate k^- divided by the binding rate k^+). Parameter A is the inverse of k^+ and is convenient for setting the speed of the negative feedback. The exponent

Table 2 Parameter values for the gonadotroph model. In the open-cell models c_{tot} is not a parameter, and in the bursting model J_{in} is not a parameter

$\sigma = 0.185$	$\bar{V}_c = 400 \text{ pL}$	$c_{tot} = 2 \ \mu\text{M}$	$V_1 = 400 \text{ aMol s}^{-1}$
$K_1 = 0.2 \ \mu\text{M}$	$L = 0.37 \text{ pL s}^{-1}$	$k_a = 0.4 \ \mu\text{M}$	$k_i = 1.0 \ \mu\text{M}$
$K_d = 0.4 \ \mu\text{M}$	$A = 2 \ \mu\text{M s}$	$V_2 = 2000 \text{ aMol s}^{-1}$	$K_2 = 0.3 \ \mu\text{M}$
$\eta = 0.01$	$J_{in} = 1200 \text{ aMol s}^{-1}$	$g_{Ca} = 20 \ \mu\text{S cm}^{-2}$	$g_K = 20 \ \mu\text{S cm}^{-2}$
$g_{SK} = 8 \ \mu\text{S cm}^{-2}$	$C_m = 1 \ \mu\text{F cm}^{-2}$	$V_3 = -3 \text{ mV}$	$V_4 = -20 \text{ mV}$
$s_1 = -20 \text{ mV}$	$s_2 = 30 \text{ mV}$	$\phi = 12 \ s^{-1}$	$K_{SK} = 0.5 \ \mu\text{M}$
$\alpha = 0.2 \ (\text{aMol} \cdot \text{cm}^2)/\text{nC}$	$P = 26,640 \text{ pL s}^{-1}$		

of 3 in Eq. 1.52 reflects the fact that the IP$_3$ receptor is a homotrimer, with three identical subunits. Finally,

$$J_{ER-out} = J_{leak} + J_{IP3}. \tag{1.56}$$

Parameter values are given in Table 2.

Summarizing, the closed-cell model consists of the two differential equations

$$\frac{dc}{dt} = (J_{ER-out} - J_{ER-in})/\bar{V}_c \tag{1.57}$$

$$\frac{dh}{dt} = (h_\infty - h)/\tau_h \tag{1.58}$$

where J_{ER-out} is given by Eq. 1.56, J_{ER-in} is given by Eq. 1.50, h_∞ is given by Eq. 1.54, and τ_h is given by Eq. 1.55.

Figure 20 shows the dynamics of the closed-cell model in response to a pulse of IP$_3$. Initially the system is at rest with c near 0 and h near 1 (the IP$_3$ receptors are not inactivated). When IP$_3$ is introduced the system produces periodic c spikes (Fig. 20A, black curve). The upstroke of each spike is due to Ca^{2+} activation of the IP$_3$ receptors and the subsequent release of Ca^{2+} from the ER into the cytosol. The downstroke of each spike is due to Ca^{2+} inactivation of the IP$_3$ receptors reflected in a decline in h (Fig. 20A, red curve). Each spike causes release of Ca^{2+} from the ER (Fig. 20B, black curve) and the ER is subsequently replenished by flux through SERCA pumps (Fig. 20B, red curve) that occurs between spikes. Thus, the total flux from the ER to the cytosol (J_{tot}) is positive during the Ca^{2+} spike and negative during the refilling stage between spikes (Fig. 20C, black curve). If the total flux is averaged over time, then one observes a rise in the time average ($< J_{tot} >$) during a spike and a slower decline during the refilling stage (Fig. 20C, red curve). When $< J_{tot} >$ reaches 0 the system has been completely reset and another spike is produced.

The level of GnRH acting on the gonadotrophs is transduced into an IP$_3$ level via the Gα_q signaling pathway, and this determines the Ca^{2+} dynamics of the cell (Stojilković et al. (1993)). This is demonstrated with the closed-cell bifurcation diagram in Fig. 21. At low IP$_3$ concentrations the system is at rest, represented

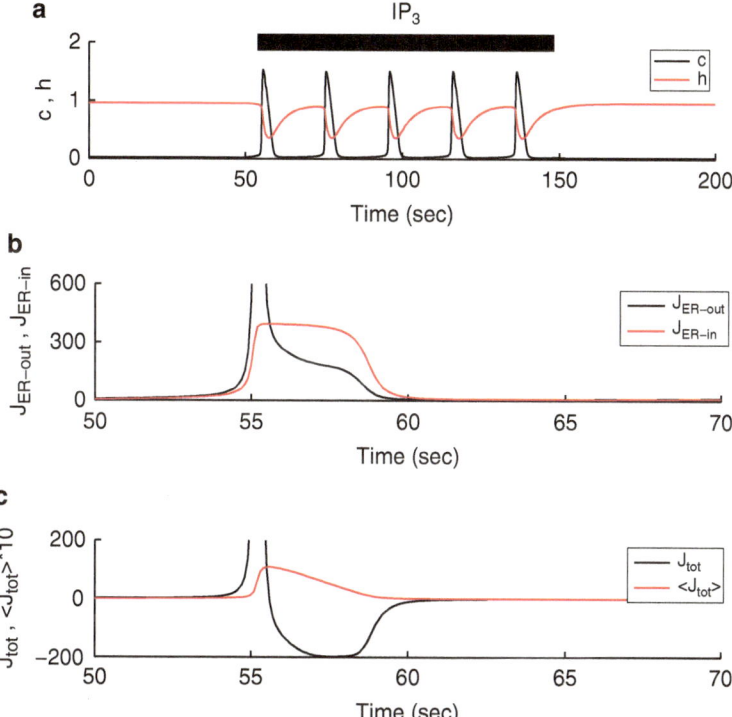

Fig. 20 Dynamics of the closed-cell gonadotroph model in response to a square pulse of IP_3 (0.8 μM). (**A**) When IP_3 is present the system produces a continuous train of Ca^{2+} spikes that are due to fast activation and slow inactivation of IP_3 receptors. (**B**) In the time frame during and after the first spike, there is an initial release of Ca^{2+} from the ER and a subsequent phase of refilling. (**C**) The initial net movement of Ca^{2+} from the ER to the cytosol is compensated by a slower replenishment of Ca^{2+} into the ER. Note the difference in time scales between the upper and lower panels

by the lower stable stationary branch of the bifurcation diagram. The stationary solution loses stability at an SNIC (Saddle-Node on Invariant Circle) bifurcation and a stable periodic solution is born. The periodic oscillations resemble "spikes" produced in neural system, and are often referred to as *calcium spikes*. As IP_3 is increased further the periodic spiking solution loses stability at a saddle-node of periodics bifurcation, and the system is attracted to a stationary solution born at a subcritical Hopf bifurcation. Thus, oscillations occur only for an intermediate range of IP_3 values. This is shown in Fig. 22 in terms of the c time courses. As IP_3 is increased in steps the system moves from a stationary, to an oscillatory, and back to a stationary state, but now with an elevated level of c. Note also that the oscillation frequency is greater at $IP_3 = 1.2\ \mu$M than at $IP_3 = 0.72\ \mu$M, as one would expect since the spiking solution is born near $IP_3 = 0.7\ \mu$M at an infinite-period SNiC bifurcation. The dynamics of a closed cell system was shown with a more detailed model in Li and Rinzel (1994) and Li et al. (1994).

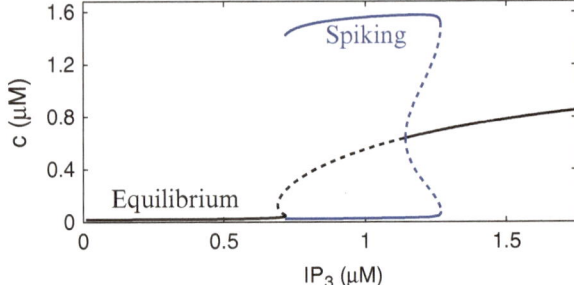

Fig. 21 Bifurcation diagram of the closed-cell gonadotroph model with IP$_3$ concentration as bifurcation parameter. The stationary solution loses stability, and an infinite-period spiking solution is born, at an SNIC bifurcation for $IP_3 = 0.7 \ \mu M$. The stationary solution regains its stability at a subcritical Hopf bifurcation ($IP_3 = 1.2 \ \mu M$), and for a small interval is bistable with the spiking solution, which ultimately disappears at a saddle-node of periodics bifurcation

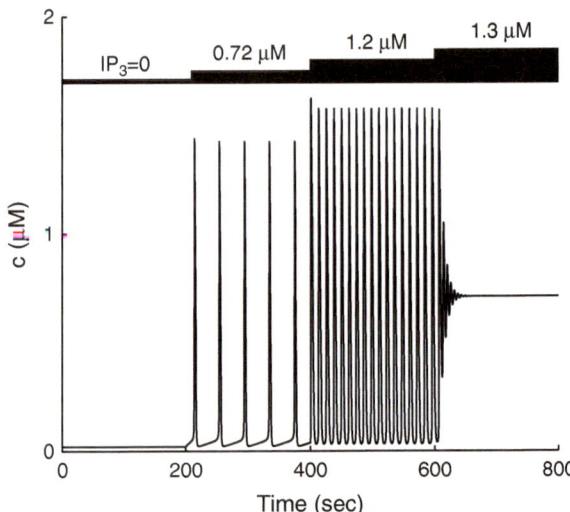

Fig. 22 The closed-cell gonadotroph model exhibits frequency modulation in response to step increases in the IP$_3$ concentration. When the concentration is too low or too high the system has a globally stable rest state

6.2 Open-Cell Dynamics

In the actual gonadotroph, Ca^{2+} enters and leaves the cell through the plasma membrane. We incorporate these fluxes next (Fig. 19B). For now we treat the Ca^{2+} influx as a parameter, J_{in},

$$J_{mem-in} = J_{in}. \tag{1.59}$$

The removal of Ca^{2+} from the cell is through plasma membrane pumps, which is modeled as a second-order Hill function, as it was with flux through SERCA pumps:

$$J_{mem-out} = \frac{V_2 c^2}{K_2^2 + c^2} \qquad (1.60)$$

where V_2 is the maximum pump rate and K_2 is the Ca^{2+} level for half maximal pumping. Adding these additional Ca^{2+} fluxes to the c equation we obtain

$$\frac{dc}{dt} = [J_{ER-out} - J_{ER-in} + \eta(J_{mem-in} - J_{mem-out})]/\bar{V}_c \qquad (1.61)$$

where η is the ratio of the plasma membrane to ER surface area. There is also a new differential equation, for the total Ca^{2+} concentration in the cell:

$$\frac{dc_{tot}}{dt} = \eta(J_{mem-in} - J_{mem-out})/\bar{V}_c. \qquad (1.62)$$

The open cell model with constant Ca^{2+} influx then consists of the three differential equations

$$\frac{dc}{dt} = [J_{ER-out} - J_{ER-in} + \eta(J_{mem-in} - J_{mem-out})]/\bar{V}_c \qquad (1.63)$$

$$\frac{dh}{dt} = (h_\infty - h)/\tau_h \qquad (1.64)$$

$$\frac{dc_{tot}}{dt} = \eta(J_{mem-in} - J_{mem-out})/\bar{V}_c \qquad (1.65)$$

where all functions other than those above are identical to those used in the closed-cell model. Parameter values are given in Table 2.

The response of the open-cell model to a pulse of IP$_3$ is shown in Fig. 23A. As with the closed-cell model, a train of Ca^{2+} spikes is produced (black curve) due to the cycling of Ca^{2+} between the cytosol and the ER. However, when the influx of Ca^{2+} into the cell is turned off, both the amplitude and the frequency of the spike train decline over time as the total Ca^{2+} concentration in the cell declines (red curve). Eventually there is not enough Ca^{2+} in the cell to sustain the oscillations and the system comes to rest at a low level of c. This simulation replicates experimental findings done in a Ca^{2+}-deficient medium (Li et al. (1994)).

The influence of the Ca^{2+} influx parameter can be described in terms of a 2-fast/1-slow analysis. Here, c and h form the fast subsystem and c_{tot} is the single slow variable. When there is sufficient Ca^{2+} flux into the cell c_{tot} is elevated, and when IP$_3$ is present the fast system exhibits stable periodic motion. Thus, there is a stable limit cycle in the h-c plane, the thick curve labeled with a green asterisk in Fig. 23B. When Ca^{2+} influx is terminated c_{tot} slowly declines, with subsequent modification of the fast-subsystem limit cycle. The limit cycle shifts slowly rightward and the

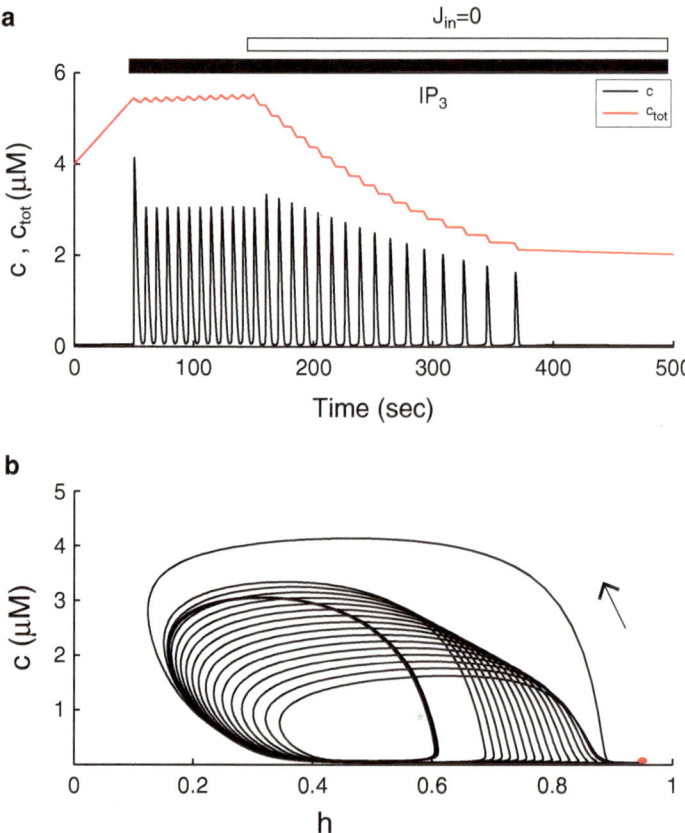

Fig. 23 The open-cell gonadotroph model with Ca^{2+} influx to the cell treated as a parameter, J_{in}. (**A**) When $J_{in} = 1200$ aMol s^{-1} and the application of IP$_3$ (0.7 μM) is simulated the system produces a train of Ca^{2+} pulses. These decline and become slower when Ca^{2+} influx is terminated, due to a slow decline in the total Ca^{2+} concentration in the cell. (**B**) In the plane of the fast variables, the decline of the slow variable c_{tot} moves the system from stable periodic motion (green asterisk) through a range of cycles of diminishing amplitude and period to a stable rest state (red circle)

amplitude in the c variable becomes smaller. When c_{tot} reaches ≈ 2 μM the rest state of the fast subsystem becomes stable (Fig. 23B, red circle). Thus, the dynamics of this open-cell model are essentially the same as those of the closed-cell model, where the parameter c_{tot} is replaced by a slowly changing variable.

This is illustrated in another way in Fig. 24. In this bifurcation diagram, IP$_3$ is held constant at 0.7 μM and c_{tot} is treated as a bifurcation parameter. Periodic spiking occurs for a large range of c_{tot} values, and is replaced by stable stationary solutions at low and high c_{tot} values. The spiking branch is born at an SNIC bifurcation, so for parameter values near here the spiking will be slow. For larger c_{tot} values the amplitude and frequency of the oscillation increase, and eventually

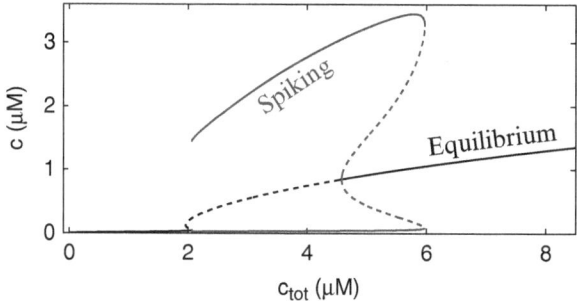

Fig. 24 Bifurcation diagram for the open-cell gonadotroph model with $IP_3 = 0.7 \ \mu M$ and c_{tot} treated as the bifurcation parameter. The periodic spiking branch is born at an SNIC bifurcation ($c_{tot} = 2.1 \ \mu M$) and disappears at a saddle-node of periodics bifurcation ($c_{tot} = 6 \ \mu M$). The stationary branch regains stability at a subcritical Hopf bifurcation ($c_{tot} = 4.6 \ \mu M$) and there is a substantial interval of bistability between periodic and stationary solutions

the periodic solution terminates at a saddle-node of periodics bifurcation. There is a fairly large interval in which the spiking and stationary solutions are both stable.

An alternate mechanism for IP$_3$-induced Ca^{2+} oscillations involves the feedback of Ca^{2+} onto IP$_3$ production or degradation. In this case, the IP$_3$ concentration itself oscillates, evoking periodic release of Ca^{2+} from the ER. While this is not the mechanism for agonist-mediated Ca^{2+} oscillations in gonadotrophs (Stojilković et al. (1993)), it can be the mechanism in other cell types (Sneyd et al. (2006)). In this report, Sneyd and colleagues describe an experimental method based on exogenous pulses of IP$_3$ for determining which type of the two feedback mechanisms for agonist-induced Ca^{2+} oscillations, called class 1 and class 2, is responsible for the observed Ca^{2+} oscillations in two cell types, pancreatic acinar cells and airway smooth muscles. Both mechanisms, Ca^{2+}-induced Ca^{2+} release and Ca^{2+}-dependent changes in IP$_3$ concentration, likely are present in a typical cell, so the real question is which, if either, mechanism dominates the Ca^{2+} dynamics.

In recent work (Harvey et al. (2010, 2011)), the authors used geometric singular perturbation theory (GSPT) to study the underlying dynamic nature in many models of Ca^{2+} dynamics. Identifying the different time scales in a given Ca^{2+} model provides a first simple diagnostic tool to predict class 1 or class 2 dominated Ca^{2+} dynamics. Furthermore, GSPT is able to explain observed anomalous delays in the Ca^{2+} dynamics (Harvey et al. (2010)) that are usually not predicted by the IP$_3$ pulse protocol experiment (Sneyd et al. (2006)). Canard theory in arbitrary dimensions (Wechselberger (2012)) is the key to explain this 'anomalous' phenomenon.

6.3 Store-Generated Bursting

We now add the final component to the model by replacing the constant Ca^{2+} influx parameter J_{in} with a term reflecting Ca^{2+} influx through Ca^{2+} ion channels

(Fig. 19C). The channels are gated by the membrane potential V, which is itself determined by ionic current through a number of ion channels. In our minimal model only three ion channel types are included. Ionic current through L-type Ca^{2+} channels, I_{Ca}, is responsible for the upstroke of an electrical action potential and also brings Ca^{2+} into the cell. It is modeled as in the lactotroph model,

$$I_{Ca}(V) = g_{Ca}m_\infty(V)(V - V_{Ca}). \tag{1.66}$$

The downstroke of an action potential is due to the slower-activating delayed-rectifying K^+ current,

$$I_K(V, n) = g_K n(V - V_K). \tag{1.67}$$

Finally, there is an SK-type K^+ current that is gated by cytosolic Ca^{2+},

$$I_{SK}(V, c) = g_{SK}s_\infty(c)(V - V_K). \tag{1.68}$$

The differential equations for the cell's electrical activity are:

$$C_m \frac{dV}{dt} = -[I_{Ca}(V) + I_K(V, n) + I_{SK}(V, c)] \tag{1.69}$$

$$\frac{dn}{dt} = \frac{n_\infty(V) - n}{\tau_n(V)}. \tag{1.70}$$

Steady state and time constant functions are:

$$m_\infty(V) = \frac{1}{2}\left(1 + \tanh\left(\frac{V - V_3}{s_1}\right)\right) \tag{1.71}$$

$$n_\infty(V) = \frac{1}{2}\left(1 + \tanh\left(\frac{V - V_4}{s_2}\right)\right) \tag{1.72}$$

$$\tau_n(V) = \left(\phi \cosh\left(\frac{V - V_4}{2s_2}\right)\right)^{-1} \tag{1.73}$$

$$s_\infty(c) = \frac{c^4}{(K_{SK}^4 + c^4)}. \tag{1.74}$$

The ordinary differential equations for the cell's electrical activity are coupled to those for Ca^{2+} handling, which again are

$$\frac{dc}{dt} = [J_{ER-out} - J_{ER-in} + \eta(J_{mem-in} - J_{mem-out})]/\bar{V}_c \tag{1.75}$$

$$\frac{dh}{dt} = (h_\infty - h)/\tau_h \tag{1.76}$$

$$\frac{dc_{tot}}{dt} = \eta(J_{mem-in} - J_{mem-out})/\bar{V}_c \qquad (1.77)$$

All functions are identical to those defined previously, except that the Ca^{2+} influx to the cell is no longer constant, but is:

$$J_{mem-in} = -\alpha I_{Ca}(V) \qquad (1.78)$$

where α converts current into Ca^{2+} ion flux. It is this term that provides the coupling from the electrical activity to the Ca^{2+} dynamics. Equation 1.68 provides the feedback from the Ca^{2+} dynamics to the electrical activity. Parameter values are given in Table 2.

The time course of the full gonadotroph model is shown in Fig. 25. In the absence of the agonist GnRH the model cell produces a tonic spiking pattern since the Ca^{2+} level in the cytosol is low, so there is very little activation of the hyperpolarizing SK current. When IP_3 is added to the model cell, simulating the effects of the physiological activator GnRH, the Ca^{2+} oscillator is activated. This is reflected in the large Ca^{2+} spikes in Fig. 25B. Each spike activates Ca^{2+}-activated SK channels, hyperpolarizing the cell for the duration of the Ca^{2+} spike. The resulting electrical bursting is then due to periodic interruptions in the tonic spiking pattern caused

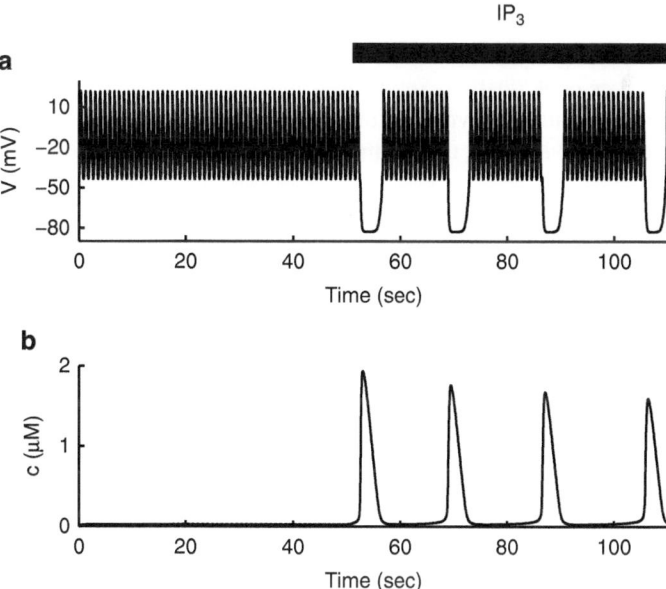

Fig. 25 The open-cell gonadotroph model with Ca^{2+} influx through voltage-dependent Ca^{2+} channels. (**A**) The tonic spiking pattern is converted to bursting when IP_3 (0.7 μM) is added to the model cell. (**B**) The cytosolic Ca^{2+} level is low until the Ca^{2+} oscillator is activated by IP_3, producing spikes of Ca^{2+} efflux from the ER into the cytosol

by the Ca^{2+} oscillator. Unlike the pseudo-plateau bursting in which c is highest during the active phase of the burst, in this store-generated bursting the peaks in the cytosolic Ca^{2+} concentration occur during the silent phase of the burst.

7 Conclusion

In this chapter we have provided examples of how mathematical modeling can be and has been used to better understand electrical and calcium dynamics in bursting pituitary cells. The difference in the bursting mechanisms of the different cell types parallels differences in secretion patterns. Lactotrophs and somatotrophs can secrete their hormones spontaneously, due to their bursting pattern that facilitates Ca^{2+} entry. Gonadotrophs, on the other hand, usually exhibit spikes that are too brief to allow much Ca^{2+} entry. Significant luteinizing hormone release only occurs when the cells are stimulated by GnRH to produce large rhythmic Ca^{2+} releases from the ER that drive secretion. Bursting oscillations also occur in some of the hypothalamic neurons that modulate the activity of pituitary cells. For example, bursting occurs in oxytocin neurons of the paraventrical nucleus during lactation, and a potential mechanism for these oscillations has been demonstrated through mathematical modeling (Rossoni et al. (2008)). Bursting also occurs in vasopressin neurons that are responsible for osmoregulation, and this too has been modeled (Clayton et al. (2010)). Bursting in hypothalamic GnRH neurons has been the focus of several modeling studies (Duan et al. (2011), Fletcher and Li (2009), LeBeau et al. (2000), Lee et al. (2010)). Yet, there is much work ahead in this field, both in understanding the intrinsic dynamics of individual cell types, and in understanding their interactions. It is likely that mathematical modeling and analysis will play a key role.

8 Appendix

Computer codes for all models discussed in this chapter are available as freeware from http://www.math.fsu.edu/~bertram/software/pituitary.

8.1 The Chay-Keizer Model

The Chay-Keizer model for bursting in the pancreatic β-cell (Chay and Keizer (1983)) was one of the first bursting models analyzed using a fast/slow analysis technique (Rinzel and Lee (1985)). A modified form of the model was used in Teka et al. (2011b) to analyze the transition between plateau and pseudo-plateau bursting, and was used for this purpose in Fig. 4. The modified Chay-Keizer model is similar in many ways to the lactotroph model used in most of this article. The differential equations are:

$$C_m \frac{dV}{dt} = -[I_{Ca}(V) + I_K(V, n) + I_{SK}(V, c) + I_{K(ATP)}(V)] \tag{1.79}$$

$$\frac{dn}{dt} = \frac{n_\infty(V) - n}{\tau_n} \tag{1.80}$$

$$\frac{dc}{dt} = -f_c(\alpha I_{Ca}(V) + k_c c). \tag{1.81}$$

where in place of the V-dependent BK K$^+$ current there is a K$^+$ current whose conductance is regulated by intracellular levels of ATP. If, as assumed here, the ATP concentration is constant, then $g_{K(ATP)}$ is a constant-conductance K$^+$ current given by

$$I_{K(ATP)}(V) = g_{K(ATP)}(V - V_K) \ . \tag{1.82}$$

The form of the steady state activation curves $m_\infty(V)$ and $n_\infty(V)$ are the same as for the lactotroph model (Eqs. 1.5, 1.7). The steady state activation function for the SK current is a third-order Hill function, rather than second-order:

$$s_\infty(c) = \frac{c^3}{c^3 + K_d^3}. \tag{1.83}$$

Parameter values for this model are given in Table 3.

8.2 The Lactotroph Model with an A-Type K$^+$ Current

An alternate lactotroph model was developed by Toporikova et al. (2008) and analyzed in Vo et al. (2010) and Vo et al. (2012). In this model, the equation for intracellular Ca^{2+} concentration is removed, as are the SK and BK currents. Instead, an A-type K$^+$ current is included that activates instantaneously and inactivates on a slower time scale. The differential equations are:

Table 3 Parameter values for the simplified Chay-Keizer model

$g_{Ca} = 1$ nS	$g_K = 2.7$ nS	$g_{SK} = 0.4$ nS	$g_{K(ATP)} = 0.18$ nS
$V_{Ca} = 25$ mV	$V_K = -75$ mV	$C_m = 5.3$ pF	$\alpha = 4.5 \times 10^{-3}$ pA$^{-1}\mu$M
$\tau_n = 18.7$ ms	$f_c = 0.00025$	$k_c = 0.5$ ms^{-1}	$K_d = 0.3\ \mu$M
$v_n = -16$ mV	$s_n = 5$ mV	$v_m = -20$ mV	$s_m = 12$ mV

Table 4 Parameter values for the lactotroph model with A-type K^+ current

$g_{Ca} = 2$ nS	$g_K = 0 - 10$ nS	$g_L = 0.3$ nS	$g_A = 4$ nS
$V_{Ca} = 50$ mV	$V_K = -75$ mV	$C_m = 2$ pF	$\tau_n = 40$ ms
$\tau_e = 20$ ms	$v_n = -5$ mV	$s_n = 10$ mV	$v_m = -20$ mV
$s_m = 12$ mV	$v_a = -20$ mV	$s_a = 10$ mV	$v_e = -60$ mV
$s_e = 10$ mV			

$$C_m \frac{dV}{dt} = -[I_{Ca}(V) + I_K(V, n) + I_A(V, e) + I_L(V)] \quad (1.84)$$

$$\frac{dn}{dt} = \frac{n_\infty(V) - n}{\tau_n} \quad (1.85)$$

$$\frac{de}{dt} = \frac{e_\infty(V) - e}{\tau_e} . \quad (1.86)$$

where the I_{Ca} and I_K currents are as before, I_L is a constant-conductance leakage current, and I_A is the A-type K^+ current:

$$I_L(V) = g_L(V - V_K) \quad (1.87)$$

$$I_A(V, e) = g_A a_\infty e(V - V_K) . \quad (1.88)$$

The activation function for I_A is

$$a_\infty(V) = \left(1 + \exp(\frac{v_a - V}{s_a})\right)^{-1} \quad (1.89)$$

and the inactivation function is

$$e_\infty(V) = \left(1 + \exp(\frac{V - v_e}{s_e})\right)^{-1} . \quad (1.90)$$

Parameter values are given in Table 4.

Acknowledgements This work was supported by NSF grants DMS 0917664 to RB, DMS 1220063 to RB and JT, and NIH grant DK 043200 to RB and JT.

References

Baer SM, Gaekel EM (2008) Slow acceleration and deacceleration through a Hopf bifurcation: Power ramps, target nucleation, and elliptic bursting. Phys Rev 78:036205

Baer SM, Erneux T, Rinzel J (1989) The slow passage through a Hopf bifurcation: Delay, memory effects, and resonance. SIAM J Appl Math 49:55–71

Benoit E (1983) Syst'emes lents-rapids dans r3 et leur canards. Asterique 109–110:159–191

Bertram R, Butte MJ, Kiemel T, Sherman A (1995) Topological and phenomenological classification of bursting oscillations. Bull Math Biol 57:413–439

Bertram R, Sherman A, Satin LS (2010) Electrical bursting, calcium oscillations, and synchronization of pancreatic islets. In: Islam MS (ed) The Islets of Langerhans, Springer, pp 261–279

Brons M, Krupa M, Wechselberger M (2006) Mixed mode oscillations due to the generalized canard phenomenon. Fields Inst Commun 49:39–63

Chay TR, Keizer J (1983) Minimal model for membrane oscillations in the pancreatic β-cell. Biophys J 42:181–190

Clayton TF, Murray AF, Leng G (2010) Modelling the *in vivo* spike activity of phasically-firing vasopressin cells. J Neuroendocrinology 22:1290–1300

Coombes S, Bressloff PC (2005) Bursting: The Genesis of Rhythm in the Nervous System. World Scientific

Crunelli V, Kelly JS, Leresche N, Pirchio M (1987) The ventral and dorsal lateral geniculate nucleus of the rat: Intracellular recordings in vitro. J Physiol 384:587–601

Dean PM, Mathews EK (1970) Glucose-induced electrical activity in pancreatic islet cells. J Physiol 210:255–264

Del Negro CA, Hsiao CF, Chandler SH, Garfinkel A (1998) Evidence for a novel bursting mechanism in rodent trigeminal neurons. Biophys J 75:174–182

Desroches M, Krauskopf B, Osinga HM (2008a) The geometry of slow manifolds near a folded node. SIAM J Appl Dyn Syst 7:1131–1162

Desroches M, Krauskopf B, Osinga HM (2008b) Mixed-mode oscillations and slow manifolds in the self-coupled FitzHugh-Nagumo system. Chaos 18:015107

Desroches M, Guckenheimer J, Krauskopf B, Kuehn C, Osinga HM, Wechselberger M (2012) Mixed-mode oscillations with multiple time scales. SIAM Rev 54:211–288

Doedel EJ (1981) AUTO: A program for the automatic bifurcation analysis of autonomous systems. Congr Numer 30:265–284

Doedel EJ, Champneys DJ, Fairgrieve TF, Kuznetov YA, Oldeman KE, Paffenroth RC, Sandstede B, Wang XJ, Zhang C (2007) AUTO-07P: Continuation and bifurcation software for ordinary differential equations Available at http://cmvl.cs.concordia.ca

Duan W, Lee K, Herbison AE, Sneyd J (2011) A mathematical model of adult GnRH neurons in mouse brain and its bifurcation analysis. J theor Biol 276:22–34

Erchova I, McGonigle DJ (2008) Rhythms in the brain: An examination of mixed mode oscillation approaches to the analysis of neurophysiological data. Chaos 18:015115

Fenichel N (1979) Genometric singular perturbation theory. J Differ Equ 31:53–98

FitzHugh R (1961) Impulses and physiological states in theoretic models of nerve membrane. Biophys J 1:445–466

Fletcher PA, Li YX (2009) An integrated model of electrical spiking, bursting, and calcium oscillations in GnRH neurons. Biophys J 96:4514–4524

Freeman ME (2006) Neuroendocrine control of the ovarian cycle of the rat. In: Neill JD (ed) Knobil and Neill's Physiology of Reproduction, 3rd edn, Elsevier, pp 2327–2388

Guckenheimer J (2008) Singular Hopf bifurcation in systems with two slow variables. SIAM J Appl Dyn Syst 7:1355–1377

Guckenheimer J, Haiduc R (2005) Canards at folded nodes. Mosc Math J 5:91–103

Guckenheimer J, Scheper C (2011) A gemometric model for mixed-mode oscillations in a chemical system. SIAM J Appl Dyn Syst 10:92–128

Harvey E, Kirk V, Osinga H, Sneyd J, Wechselberger M (2010) Understanding anomalous delays in a model of intracelular calcium dynamics. Chaos 20:045104

Harvey E, Kirk V, Sneyd J, Wechselberger M (2011) Multiple time scales, mixed-mode oscillations and canards in models of intracellular calcium dynamics. J Nonlinear Sci 21:639–683

Hodgkin AL, Huxley AF (1952) A quantitative description of membrane current and its application to conduction and excitation in nerve. J Physiol 117:500–544

Izhikevich EM (2007) Dynamical Systems in Neuroscience. MIT Press

Keener K, Sneyd J (2008) Mathematical Physiology, 2nd edn. Springer

Krupa M, Wechselberger M (2010) Local analysis near a folded saddle-node singularity. J Differ Equ 248:2841–2888

Kukuljan M, Rojas E, Catt KJ, Stojilković SS (1994) Membrane potential regulates inositol 1,4,5-trisphosphate-controlled cytoplasmic Ca^{2+} oscillations in pituitary gonadotrophs. J Biol Chem 269:4860–4865

Kuryshev YA, Childs GV, Ritchie AK (1996) Corticotropin-releasing hormone stimulates Ca^{2+} entry through L- and P-type Ca^{2+} channels in rat corticotropes. Endocrinology 137:2269–2277

LeBeau AP, van Goor F, Stojilković SS, Sherman A (2000) Modeling of membrane excitability in gonadotropin-releasing hormone-secreting hypothalamic neurons regulated by Ca^{2+}-mobilizing and adenylyl cyclase-coupled receptors. J Neurosci 20:9290–9297

Lee K, Duan W, Sneyd J, Herbison AE (2010) Two slow calcium-activated afterhyperpolarization currents control burst firing dynamics in gonadotropin-releasing hormone neurons. J Neurosci 30:6214–6224

Li YX, Rinzel J (1994) Equations for InsP3 receptor-mediated $[Ca^{2+}]$ oscillations derived from a detailed kinetic model: a Hodgkin-Huxley like formalism. J theor Biol 166:461–473

Li YX, Rinzel J, Keizer J, Stojilković SS (1994) Calcium oscillations in pituitary gonadotrophs: Comparison of experiment and theory. Proc Natl Acad Sci USA 91:58–62

Li YX, Keizer J, Stojilković SS, Rinzel J (1995) Ca^{2+} excitability of the ER membrane: An explanation for IP3-induced Ca^{2+} oscillations. Am J Physiol 269:C1079–C1092

Lyons DJ, Horjales-Araujo E, Broberger C (2010) Synchronized network oscillations in rat tuberoinfundibular dopamine neurons: Switch to tonic discharge by thyrotropin-releasing hormone. Neuron 65:217–229

Milescu LS, Yamanishi T, Ptak K, Mogri MZ, Smith JC (2008) Real-time kinetic modeling of voltage-gated ion channels using dynamic clamp. Biophys J 95:66–87

Milik A, Szmolyan P (2001) Multiple time scales and canards in a chemical oscillator. In: Jones C, Khibnik A (eds) Multiple-Time-Scale Dynamical Systems, Springer-Verlag, IMA Vol. Math. Appl., vol 122, pp 117–140

Morris C, Lecar H (1981) Voltage oscillations in the barnacle giant muscle fiber. Biophys J 35: 193–213

Nowacki J, Mazlan S, Osinga HM, Tsaneva-Atanasova K (2010) The role of large-conductance calcium-activated K^{+} (BK) channels in shaping bursting oscillations of a somatotroph cell model. Physica D 239:485–493

Nunemaker CS, DeFazio RA, Moenter SM (2001) Estradiol-sensitive afferents modulate long-term episodic firing patterns of GnRH neurons. Endocrinology 143:2284–2292

Osinga HM, Sherman A, Tsaneva-Atanasova K (2012) Cross-currents between biology and mathematics: The codimension of pseudo-plateau bursting. Discret Contin Dyn S 32:2853–2877

Rinzel J (1987) A formal classification of bursting mechanisms in excitable systems. In: Teramoto E, Yamaguti M (eds) Lecture Notes in Biomathematics, vol 71, Springer, pp 267–281

Rinzel J, Lee YS (1985) On different mechanisms for membrane potential bursting. In: Othmer HG (ed) Nonlinear Oscilations in Biology, vol 66, Springer-Verlag, pp 19–33

Rinzel J, Keizer J, Li YX (1996) Modeling plasma membrane and endoplasmic reticulum excitability in pituitary cells. Trends Endocrinol Metab 7:388–393

Rossoni E, Feng J, Tirozzi B, Brown D, Leng G, Moos F (2008) Emergent synchronous bursting of oxytocin neuronal network. PLoS Comp Biol 4(7):1000123

Rubin J, Wechselberger M (2007) Giant squid-hidden canard: The 3D geometry of the Hodgkin-Huxley model. Biol Cybern 97:5–32

Rubin J, Wechselberger M (2008) The selection of mixed-mode oscillations in a Hodgkin-Huxley model with multiple timescales. Chaos 18:015105

Shangold GA, Murphy SN, Miller RJ (1988) Gonadotropin-releasing hormone-induced Ca^{2+} transients in single identified gonadotropes require both intracellular Ca^{2+} mobilization and Ca^{2+} influx. Proc Natl Acad Sci USA 85:6566–6570

Sharp AA, O'Neil MB, Abbott LF, Marder E (1993) Dynamic clamp–computer-generated conductances in real neurons. J Neurophysiol 69:992–995

Sherman A, Keizer J, Rinzel J (1990) Domain model for Ca^{2+}-inactivation of Ca^{2+} channels at low channel density. Biophys J 58:985–995

Sherman A, Li YX, Keizer JE (2002) Whole-cell models. In: Fall CP, Marland ES, Wagner JM, Tyson JJ (eds) Computational Cell Biology, 1st edn, Springer, pp 101–139

Sneyd J, Tsaneva-Atanasova K, Bruce JIE, Straub SV, Giovannucci DR, Yule DI (2003) A model of calcium waves in pancreatic and parotid acinar cells. Biophys J 85:1392–1405

Sneyd J, Tsaneva-Atanasova K, Reznikov V, Sanderson MJ, Yule DI (2006) A method for determining the dependence of calcium oscillations on inositol trisphosphate oscillations. Proc Natl Acad Sci USA 103:1675–1680

Stern JV, Osinga HM, LeBeau A, Sherman A (2008) Resetting behavior in a model of bursting in secretory pituitary cells: Distinguishing plateaus from pseudo-plateaus. Bull Math Biol 70:68–88

Stojilković SS, Tomić M (1996) GnRH-induced calcium and current oscillations in gonadotrophs. Trends Endocrinol Metab 7:379–384

Stojilković SS, Kukuljan M, Iida T, Rojas E, Catt KJ (1992) Integration of cytoplasmic calcium and membrane potential oscillations maintains calcium signaling in pituitary gonadotrophs. Proc Natl Acad Sci USA 89:4081–4085

Stojilković SS, Kukuljan M, Tomić M, Rojas E, Catt KJ (1993) Mechanism of agonist-induced $[Ca^{2+}]_i$ oscillations in pituitary gonadotrophs. J Biol Chem 268:7713–7720

Stojilković SS, Tabak J, Bertram R (2010) Ion channels and signaling in the pituitary gland. Endocr Rev 31:845–915

Szmolyan P, Wechselberger M (2001) Canards in $\mathbb{R}^3$. J Diff Eq 177:419–453

Szmolyan P, Wechselberger M (2004) Relaxation oscillations in $\mathbb{R}^3$. J Diff Eq 200:69–144

Tabak J, Toporikova N, Freeman ME, Bertram R (2007) Low dose of dopamine may stimulate prolactin secretion by increasing fast potassium currents. J Comput Neurosci 22:211–222

Tabak J, Tomaiuolo M, Gonzalez-Iglesias AE, Milescu LS, Bertram R (2011) Fast-activating voltage- and calcium-dependent potassium (BK) conductance promotes bursting in pituitary cells: A dynamic clamp study. J Neurosci 31:16,855–16,863

Teka W, Tabak J, Vo T, Wechselberger M, Bertram R (2011a) The dynamics underlying pseudo-plateau bursting in a pituitary cell model. J Math Neurosci 1:12, DOI 10.1186/2190-8567-1-12

Teka W, Tsaneva-Atanasova K, Bertram R, Tabak J (2011b) From plateau to pseudo-plateau bursting: Making the transition. Bull Math Biol 73:1292–1311

Teka W, Tabak J, Bertram R (2012) The relationship between two fast-slow analysis techniques for bursting oscillations. Chaos 22, DOI 10.1063/1.4766943

Tomaiuolo M, Bertram R, Leng G, Tabak J (2012) Models of electrical activity: calibration and prediction testing on the same cell. Biophys J 103:2021–2032

Toporikova N, Tabak J, Freeman ME, Bertram R (2008) A-type K^+ current can act as a trigger for bursting in the absence of a slow variable. Neural Comput 20:436–451

Tsaneva-Atanasova K, Sherman A, Van Goor F, Stojilković SS (2007) Mechanism of spontaneous and receptor-controlled electrical activity in pituitary somatotrophs: Experiments and theory. J Neurophysiol 98:131–144

Tse A, Hille B (1992) GnRH-induced Ca^{2+} oscillations and rhythmic hyperpolarizations of pituitary gonadotropes. Science 255:462–464

Tse FW, Tse A, Hille B (1994) Cyclic Ca^{2+} changes in intracellular stores of gonadotropes during gonadotropin-releasing hormone-stimulated Ca^{2+} oscillations. Proc Natl Acad Sci USA 91:9750–9754

Tse FW, Tse A, Hille B, Horstmann H, Almers W (1997) Local Ca^{2+} release from internal stores controls exocytosis in pituitary gonadotrophs. Neuron 18:121–132

Van Goor F, Li YX, Stojilković SS (2001a) Paradoxical role of large-conductance calcium-activated K^+ (BK) channels in controlling action potential-driven Ca^{2+} entry in anterior pituitary cells. J Neurosci 21:5902–5915

Van Goor F, Zivadinovic D, Martinez-Fuentes AJ, Stojilković SS (2001b) Dependence of pituitary hormone secretion on the pattern of spontaneous voltage-gated calcium influx. Cell-type specific action potential secretion coupling. J Biol Chem 276:33,840–33,846

Vo T, Bertram R, Tabak J, Wechselberger M (2010) Mixed mode oscillations as a mechanism for pseudo-plateau bursting. J Comput Neurosci 28:443–458

Vo T, Bertram R, Wechselberger M (2012) Bifurcations of canard-induced mixed mode oscillations in a pituitary lactotroph model. Discret Contin Dyn S 32:2879–2912

Wechselberger M (2005) Existence and bifurcation of canards in $\mathbb{R}^3$ in the case of a folded node. SIAM J Dyn Syst 4:101–139

Wechselberger M (2012) A propos de canards (apropos canards). Trans Am Math Sci 364: 3289–3309

Wechselberger M, Weckesser W (2009) Bifurcations of mixed-mode oscillations in a stellate cell model. Physica D 238:1598–1614

Chapter 2
The Nonlinear Dynamics of Calcium

Vivien Kirk and James Sneyd

Abstract Oscillations and travelling waves in the concentration of free cytosolic calcium are complex dynamical phenomena that play vital roles in cellular function, controlling such processes as contraction, secretion and differentiation. Although, nowadays, these oscillations and waves may be observed experimentally with relative ease, we still lack a rigorous understanding of, firstly, the mechanisms underlying these waves and oscillations in different cell types, and, secondly, the mathematical structures that underlie these complex dynamics. Thus, the study of calcium waves and oscillations is one area in which modellers have, over the years, played a major role. Here, we review our current understanding of the nonlinear dynamics of calcium waves and oscillations, restricting our attention almost wholly to deterministic models.

1 Introduction

In almost every cell type, the concentration of free cytosolic calcium, $[Ca^{2+}]$, plays a major role in cellular function and regulation [5, 4]. In all muscle cells, for example, a rise in $[Ca^{2+}]$ is the signal that causes contraction [8, 40]. In cardiac and skeletal muscle, this rise in $[Ca^{2+}]$ comes about as Ca^{2+} enters the cell through voltage-gated channels in the cell membrane. The resultant high $[Ca^{2+}]$ causes myosin to bind to actin, thus exerting a contractile force. In synapses, where one neuron communicates with another, the release of neurotransmitter is governed by the $[Ca^{2+}]$ in the presynaptic terminal [96, 99], while in a completely different cell type, the parotid acinar cell (a type of epithelial cell), a rise in $[Ca^{2+}]$ causes water secretion and thus the production of saliva [1, 155].

In many cell types (hepatocytes, for example) the exact role played by Ca^{2+} is not well understood, although it seems clear that it is important for cell function,

V. Kirk • J. Sneyd (✉)

Department of Mathematics, University of Auckland, Auckland, New Zealand

e-mail: v.kirk@auckland.ac.nz; sneyd@math.auckland.ac.nz

© Springer International Publishing Switzerland 2015

R. Bertram et al., *Mathematical Analysis of Complex Cellular Activity*, Frontiers in Applied Dynamical Systems: Reviews and Tutorials 1, DOI 10.1007/978-3-319-18114-1_2

while in other cell types (such as neuroendocrine cells like gonadotropin-releasing hormone neurons) a rise in $[Ca^{2+}]$ is doubtless closely linked to the secretion of hormone, but we do not understand exactly how this link works [71].

Over the last few decades highly sophisticated methods have been developed to measure $[Ca^{2+}]$ in cells (often still situated in living animals) both in space and in time. The most important method is undoubtedly fluorescence microscopy. Investigators have developed molecules that emit light when they bind Ca^{2+}. By loading cells with such Ca^{2+} fluorescent dyes one is now able directly to observe the Ca^{2+} in the cell and show the results as a video, for example.

One of the first things that investigators noticed was that, in many cell types, the Ca^{2+} transients, far from being a simple rise and fall, have complex dynamic behaviour. In some cells, $[Ca^{2+}]$ oscillates with a period ranging from under a second to many minutes. In other, larger, cells, these oscillations are organised into periodic waves that travel at around 15 $\mu m\,s^{-1}$. In even larger cells, such as *Xenopus* oocytes, these periodic waves are organised into periodic spirals or target patterns.

According to current dogma, oscillations and periodic waves of Ca^{2+} control cellular functions in a frequency-dependent manner. Ca^{2+} itself is toxic to cells — prolonged high $[Ca^{2+}]$ will kill a cell — and thus an amplitude-modulated signal is of less use. However, by modulating the frequency of the oscillation, the signal can be carried efficiently, without endangering the cell. Although this is a useful working hypothesis, and is supported by a great deal of experimental evidence, in some cell types the actual situation is considerably more complicated, with both amplitude and frequency playing major roles, while in yet other cell types, the frequency of the oscillation appears to play almost no role at all. Examples of these different situations are discussed in more detail below, and in Fig. 1.

Three examples of Ca^{2+} oscillations and waves, from three very different cell types, are shown in Fig. 1. In the first example we believe we know what the Ca^{2+} oscillations are doing, and how their function is controlled by their frequency; in the second example, we believe we know what the Ca^{2+} oscillations are doing, but it seems that the oscillation frequency is entirely unimportant; in the third example, we think we know what the Ca^{2+} transients are doing (at least in general terms), but we don't really know how they do it.

In part A of Fig. 1 we show Ca^{2+} oscillations in airway smooth muscle cells, in response to the agonist methylcholine. These Ca^{2+} oscillations (indirectly, but through a well-known pathway [73]) cause binding of the contractile proteins, myosin and actin, and thus cause contraction of the muscle [113]. Since airway smooth muscle surrounds the airways, its contraction causes contraction of the airways and restriction of breathing. (Interestingly, there is no other known phys-iological function of airway smooth muscle; it is the only known organ whose sole function is pathological.) The extent of the muscle contraction is closely correlated with the frequency of the Ca^{2+} oscillation, and thus we believe we understand the physiological function of these oscillations. Although we call them Ca^{2+} oscillations, they are, in fact, periodic waves, as can be seen from the more detailed plot in part B. In this space-time diagram, a higher $[Ca^{2+}]$ is denoted by a

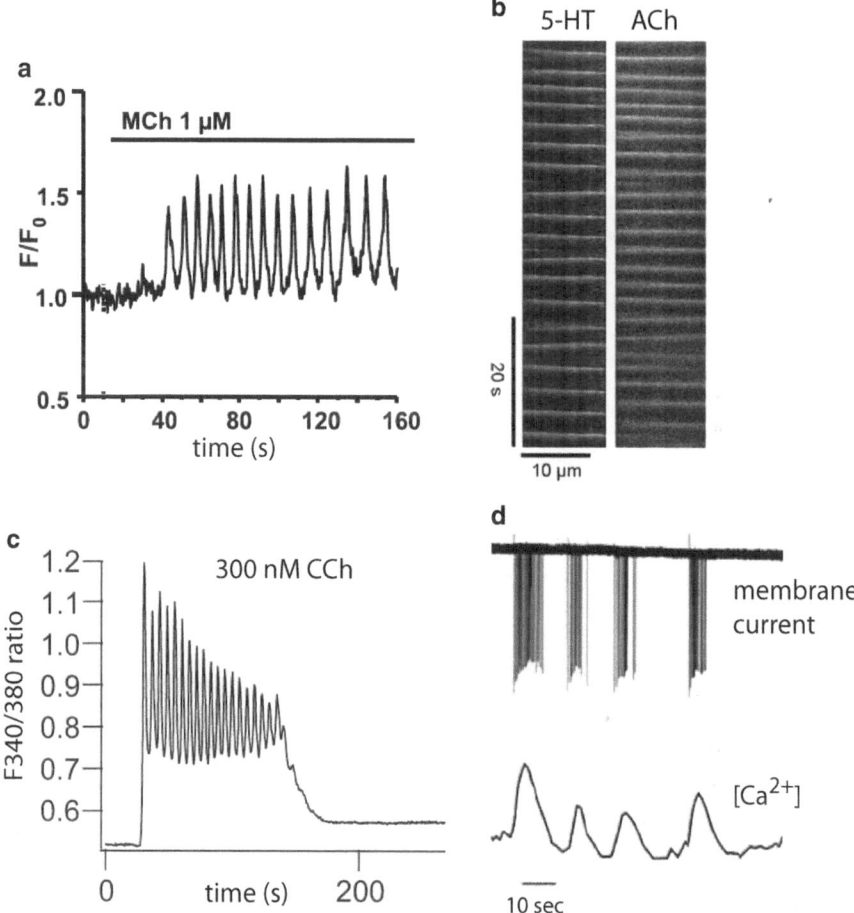

Fig. 1 Three examples of the complex behaviour of $[Ca^{2+}]$. A: Oscillations of $[Ca^{2+}]$ in human airway smooth muscle cells, in response to the agonist methylcholine (MCh). Figure modified from [107]. B: Ca^{2+} oscillations in airway smooth muscle cells, plotted in both space and time, showing that the oscillations shown in A are in fact periodic waves. Figure modified from [104]. Responses to two different agonists – serotonin (5-HT) and acetylcholine (ACh) – are shown. C: Ca^{2+} oscillations in parotid acinar cells, in response to carbochol (CCh). Figure modified from [53]. D: Ca^{2+} transients in mouse gonadotropin-releasing hormone (GnRH) neurons. Figure modified from [84]. The membrane current (upper trace) and the $[Ca^{2+}]$ concentration (lower trace) were measured simultaneously. It can be seen that each burst of electrical spikes corresponds to a transient in $[Ca^{2+}]$

lighter shade; the fact that white bands extend across the domain at a slight angle means that the Ca^{2+} oscillations are propagating across the cell to form periodic waves.

In part C of Fig. 1 we show Ca^{2+} oscillations from parotid acinar cells. The parotid gland is one of the saliva-producing glands, and parotid acinar cells are epithelial cells specialised for the transport of water. Each rise in $[Ca^{2+}]$ causes the opening of Ca^{2+}-dependent K^+ channels at one end of the cell, the opening of Ca^{2+}-dependent Cl^- channels at the other end of the cell, and thus transcellular ion flow, with water following by osmosis. However, although it was thought for some years that the rate of water flow was controlled by the frequency of the oscillation, this is now thought not to be the case [101, 102]. In this cell type the rate of water transport is governed almost entirely by the average $[Ca^{2+}]$, with the frequency of the oscillation playing no important role that we can discern.

Our final example, in part D of Fig. 1, is from a gonadotropin-releasing hormone neuron, a neuroendocrine cell in the hypothalamus that secretes gonadotropin-releasing hormone, or GnRH. The upper trace is the membrane current, which shows clear groups of electrical spikes, usually called electrical bursting. The lower trace shows the associated Ca^{2+} transient. Periodicity is not clear, so we do not call these Ca^{2+} oscillations. Although we know that these Ca^{2+} transients are associated with the secretion of GnRH, we do not understand exactly how. The secretion of GnRH appears to be controlled on a time scale of hours, while these Ca^{2+} transients occur with much shorter period, on the order of tens of seconds. How the fast Ca^{2+} transients are connected to the slow control of GnRH secretion is one of the great puzzles in the study of neuroendocrine cells.

It is clear from even this highly selective set of examples that the study of Ca^{2+} oscillations has a great deal to offer the mathematical modeller. Such complex dynamic phenomena simply cannot be properly understood without detailed quantitative models, and without a detailed understanding of the mechanisms that can drive periodic behaviour. Because of this, mathematical modellers have often played significant roles in the study of Ca^{2+} dynamics [38, 39, 44, 114, 136].

1.1 Some background physiology

Although it is not the purpose of this article to present a detailed discussion of Ca^{2+} physiology, some details are necessary in order to understand how models are constructed.

Because high $[Ca^{2+}]$ is toxic, all cells spend a great deal of energy ensuring that $[Ca^{2+}]$ is kept low. This is not an easy job, energetically speaking. All cells are bathed in a Ca^{2+}-rich environment, with a concentration of approximately 1 mM, kept at this level by continual release from the bones. However, inside the cell cytoplasm, energy-consuming pumps are continually removing Ca^{2+} to keep $[Ca^{2+}] \approx 50$ nM, about 20,000 times lower than outside the cell. There is thus an enormous concentration gradient from the outside to the inside of the cell. Hence, cells can raise $[Ca^{2+}]$ quickly, merely by opening Ca^{2+} channels in the cell membrane, but must continually expend energy to maintain this concentration gradient.

Internal cellular compartments, such as the endoplasmic (or, in muscle cells, the sarcoplasmic) reticulum (ER or SR) are also major Ca^{2+} stores, with Ca^{2+} pumps, called SERCA pumps (Sarcoplasmic/Endoplasmic Reticulum Calcium ATPases) continually pumping Ca^{2+} from the cytoplasm into the ER or SR. Similarly, the mitochondria constitute another major internal Ca^{2+} store.

Thus a cell at rest is continually expending large amounts of energy, merely to keep $[Ca^{2+}]$ low, and there is a continual low-level cycling of Ca^{2+} into and out of the cytoplasm, as Ca^{2+} leaks in, and is then removed by the pumps.

As an additional control for $[Ca^{2+}]$, of every 1000 Ca^{2+} ions entering the cytoplasm, approximately 999 are quickly bound to large proteins, called Ca^{2+} buffers, thus preventing the Ca^{2+} from harming the cell. This so-called Ca^{2+} buffering can play a major role in quantitative models (although it can have less effect on the qualitative dynamics) and often needs to be considered carefully.

To construct a model of Ca^{2+} dynamics one writes down a conservation equation that keeps track of all the Ca^{2+} entering and leaving the cytoplasm. There are a number of such Ca^{2+} fluxes (some of which, but not all, are summarised in Fig. 2).

- Ca^{2+} can flow into the cell from outside through a number of types of channel.

  - Voltage-gated Ca^{2+} channels open in response to an increase in the potential difference across the cell membrane. The resultant influx of Ca^{2+} will lead to further depolarisation and possibly to an action potential if the cell is electrically excitable.
  - Receptor-operated channels open in response (possibly quite indirectly) to the binding of agonist to a cell membrane receptor.
  - Store-operated channels open in response to a severe depletion of the ER or SR.

- Ca^{2+} is moved from the cytoplasm to outside the cell by the action of Ca^{2+} ATPase pumps in the cell membrane. Other ways in which Ca^{2+} is removed from the cytoplasm — for example, by a Na/Ca exchanger — are important in some cell types.
- Release of Ca^{2+} from the ER or SR occurs through two major channels.

  - When an agonist binds to a receptor on the cell membrane it initiates a series of reactions that ends in the production of inositol trisphosphate (IP_3), which diffuses through the cytoplasm and binds to IP_3 receptors (IPR) located on the membrane of the ER or SR. IPR are also Ca^{2+} channels, and when IP_3 binds they open, and release Ca^{2+} from the ER. Both Ca^{2+} and IP_3 modulate the open probability of the IPR. IPR exhibit the important property of Ca^{2+}-induced Ca^{2+} release, or CICR, whereupon a small increase in $[Ca^{2+}]$ leads to the opening of the IPR and the further release of Ca^{2+}. Thus, CICR is a positive feedback process in which Ca^{2+} stimulates its own release. In addition, a high $[Ca^{2+}]$ will close the IPR.
  - Ryanodine receptors (RyR) are similar to IPR, and are almost as ubiquitous. They are not opened by IP_3, but their open probability is modulated by Ca^{2+} in a manner similar to IPR. RyR also exhibit CICR, and indeed were the original

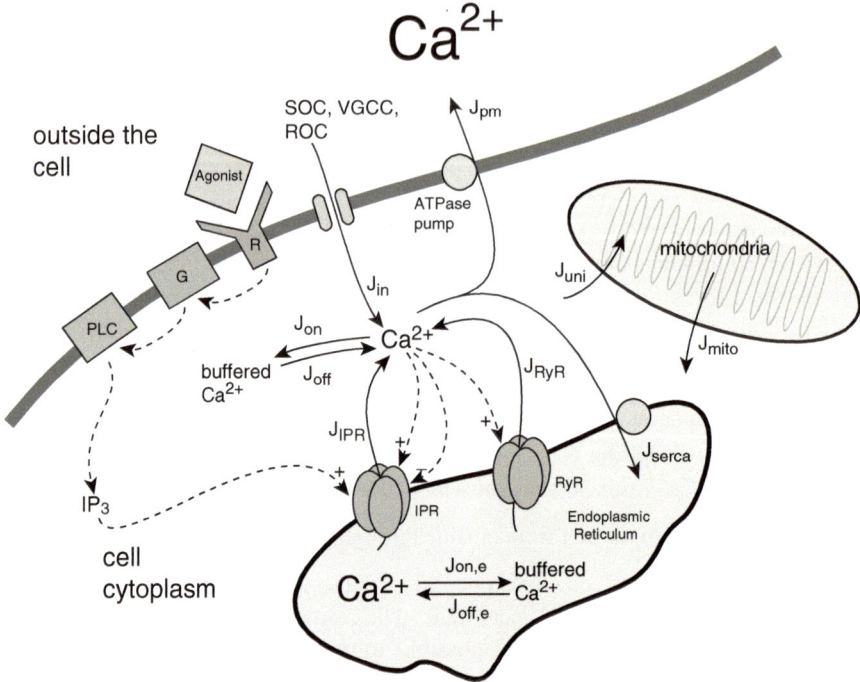

Fig. 2 Diagram of the major fluxes involved in the control of $[Ca^{2+}]$. Binding of agonist to a cell membrane receptor (R) leads to the activation of a G-protein (G), and subsequent activation of phospholipase C (PLC). This cleaves phosphotidylinositol bisphosphate into diacylglycerol and inositol trisphosphate (IP_3), which is free to diffuse through the cell cytoplasm. When IP_3 binds to an IP_3 receptor (IPR) on the endoplasmic reticulum (ER) membrane it causes the release of Ca^{2+} from the ER, and this Ca^{2+} in turn modulates the open probability of the IPR and ryanodine receptors (RyR). Calcium fluxes are denoted by solid arrows. Calcium can be released from the ER through IPR (J_{IPR}) or RyR (J_{RyR}), can be pumped from the cytoplasm into the ER (J_{serca}) or to the outside (J_{pm}), can be taken up into (J_{uni}), or released from (J_{mito}), the mitochondria, and can be bound to (J_{on}), or released from (J_{off}), Ca^{2+} buffers. Entry from the outside (J_{in}) is controlled by a variety of possible channels, including store-operated channels (SOC), voltage-gated calcium channels (VGCC), and receptor-operated channels (ROC)

type of Ca^{2+} channel in which this behaviour was discovered [41]. RyR are the predominant Ca^{2+} release channels in skeletal and cardiac muscle.

- Reuptake of Ca^{2+} into the ER/SR is done by SERCA pumps, which use the energy of ATP to pump Ca^{2+} up its concentration gradient.
- There are also important Ca^{2+} fluxes to and from the mitochondria. However, we shall not be considering such fluxes in detail here, as they tend to play less important roles in many current models of Ca^{2+} dynamics. As always, there are multiple exceptions to this rule [23, 24, 31, 45, 59, 91, 92, 95, 106].

Given these fluxes, one possible mechanism of Ca^{2+} oscillations becomes a little clearer. When an agonist binds to its receptor it begins the process that results in the production of IP_3. This initiates an explosive release of Ca^{2+} from the ER/SR, via a process of CICR. Once $[Ca^{2+}]$ is high enough the IPR shuts and Ca^{2+} efflux from the ER/SR is terminated. As long as the IPR enters a refractory state, thus preventing immediate reopening, Ca^{2+} pumps can remove Ca^{2+} from the cytoplasm and the cycle can repeat. A similar process occurs through the RyR also, and in many cases both IPR and RyR collaborate to produce the oscillations [133, 146, 149].

It is important to note that there are some cell types, most notably skeletal and cardiac muscle, in which CICR is crucial for cellular function, but does not result in sustained Ca^{2+} oscillations. In skeletal and cardiac muscle, the entry of a small amount of Ca^{2+} through voltage-gated channels (in response to electrical depolarisation) initiates CICR through RyR, which releases a large amount of Ca^{2+} into the cytoplasm, activating the contractile machinery and leading to contraction of the cell. However, each Ca^{2+} transient is caused by an action potential which is generated elsewhere – for cardiac cells this is the sino-atrial node – and thus the muscle cell itself exhibits no intrinsic oscillatory behaviour, at least under normal conditions. It is possible to force cardiac cells into a regime where the ER is overloaded with Ca^{2+}, and will thus generate spontaneous rhythmic Ca^{2+} transients, but this is pathological behaviour. For this reason we shall spend less time here considering Ca^{2+} dynamics in cardiac and skeletal muscle. Interested readers are referred to the comprehensive reviews of [8, 40].

We have described above one possible mechanism that can cause Ca^{2+} oscillations. However, there are many others [30, 44, 73, 114]. For example, Ca^{2+} can affect the production and the degradation of IP_3, forming both positive and negative feedback loops which are theoretically capable [37, 105] of generating oscillations (Fig. 3).

It is very important to understand that, although Ca^{2+} oscillations may look quite similar in different cell types, with similar periods and shapes, such similarity in appearance can be quite deceptive. Different cell types can, and in general do, have quite different mechanisms generating their Ca^{2+} oscillations, and it is unwise to extrapolate mechanisms from one cell type to another, based solely on a desire for simplicity and a fortuitous convergence of appearance. Thus, although the basic toolbox (see section 2) is the same from one cell to another, the way in which those tools are combined and used can be quite different, and each cell must be treated on its own merits.

When Ca^{2+} release occurs in a particular part of the cytoplasm, Ca^{2+} can diffuse to neighbouring release sites (either IPR or RyR) and initiate further release of Ca^{2+} there, thus propagating a travelling wave of increased $[Ca^{2+}]$. In such a way are oscillations converted to periodic waves. These waves travel at approximately 10–15 μms^{-1} and, in larger cell types such as the *Xenopus* oocyte, can form spiral waves and target patterns [83]. Calcium waves can also travel between cells, in regions extending over many cells [85], although this review shall not discuss such intercellular waves at all.

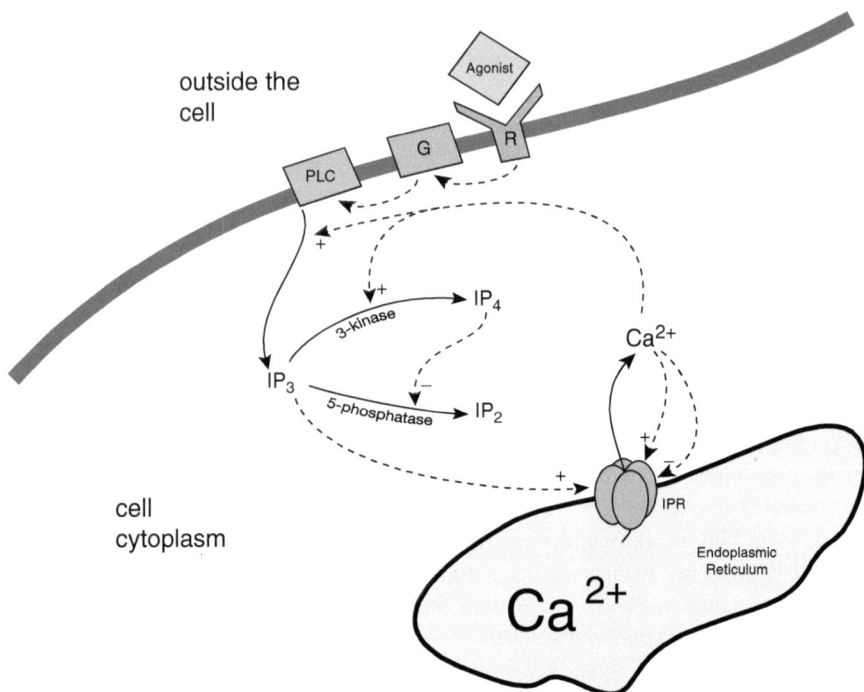

Fig. 3 Schematic diagram of some of the interactions between Ca^{2+} and IP_3. Calcium can activate PLC, leading to an increase in the rate of production of IP_3, and it can also increase the rate at which IP_3 is phosphorylated by the 3-kinase. The end product of phosphorylation by the 3-kinase, IP_4, acts as a competitive inhibitor of dephosphorylation of IP_3 by the 5-phosphatase. Not all of these feedbacks are significant in every cell type

It is not the purpose of this review to enumerate and discuss all the possible ways in which Ca^{2+} oscillations and waves are thought to arise in different cell types, as this would be a Herculean task. Instead we shall focus on a mathematical analysis of a few of the major mechanisms. The techniques we discuss here will be equally applicable to all the other oscillatory mechanisms and models.

The variety of mechanisms underlying Ca^{2+} oscillations and waves is matched by their variety of physiological function. We have already seen specific examples of how Ca^{2+} oscillations control the contraction of smooth muscle, the transport of water by exocrine epithelia and the secretion of hormones. However, Ca^{2+} oscillations are also known to control fertilisation, proliferation, cell metabolism, vesicle secretion, and even information processing in neurons. Again, we shall not discuss such matters in this review, but refer instead to the many excellent reviews on the topic [4, 5, 39, 44].

1.2 Overview of calcium models

There are two major types of model of Ca^{2+} dynamics: the spatially homogeneous model, which assumes a well-mixed cell and uses ordinary differential equations, and the spatially inhomogeneous model, which allows for spatial variation of $[Ca^{2+}]$ and uses partial differential equations (usually a reaction-diffusion equation). Within each of these divisions, models can be deterministic or stochastic, and can be essentially arbitrarily complex. PDE models, in particular, can become extremely complex, with microdomains of Ca^{2+}, i.e., small localised regions where, because of geometric restrictions, the Ca^{2+} concentration is orders of magnitude higher than in other parts of the cell.

It is important to note that the type of model one constructs is not essentially dependent on what is believed to be the "real" situation. For example, it is perfectly well known that cells are not well mixed, and that $[Ca^{2+}]$ is not even close to homogeneous. Nevertheless, a well-mixed model can still be a useful tool, guiding new experimental results and making testable predictions. Similarly, we know also that, at the highest level of detail, the release of Ca^{2+} through either IPR or RyR is inherently a stochastic, not a deterministic process. In some situations this matters, and stochastic models must be used. In other cases, stochastic aspects are less important.

In other words, we construct models, not to be the most detailed and accurate representation of what we believe is the true situation, but to be useful tools to guide our understanding. Depending on what we wish to understand, we construct a model to suit our needs. This is something that is worth emphasising. It is not uncommon for models to be criticised for omitting aspects that exist in the real cell. Since it is hardly possible for models to do otherwise, such criticisms are facile. What really matters is whether or not the model contains the mechanisms that matter for the particular question under investigation.

Conversely, modellers commonly make an analogous mistake; often they construct a model, show that some solutions look the same as experiments, and claim success. This is, of course, equally as facile as the criticisms mentioned above. A similarity of appearance is rarely a guide to underlying mechanism. It is not until the model is used as a predictive tool, and not until experiments are done to test these model predictions, that a model is useful. It matters not whether the experiments confirm or reject the model predictions. The important thing is that the model has been used to advance our understanding.

Whether the model consists of ODEs or PDEs, the basic approach is similar. There are certain cellular components which tend to be common across all cell types, and have reasonably standard models. For instance, the SERCA pumps that transport Ca^{2+} from the cytoplasm, up its concentration gradient into the ER or SR, are ubiquitous, and tend always to be modelled in similar ways. Similarly, there are voltage-gated Ca^{2+} channels, IPR and RyR, Ca^{2+} buffers, and various other Ca^{2+} channels, pumps and exchangers, each of which tends to come with a relatively well-accepted model.

Thus, one useful concept is that of a Ca^{2+} "toolbox" [6]. This toolbox contains a variety of Ca^{2+} transport mechanisms, or modules, from which we can select the most appropriate to build a model in any particular situation. The question of model construction then comes down, in essence, to selection of which modules are the best to use (given the question under consideration), and which is the best model to use for each module. Of course, since there are a very large number of modules in our Ca^{2+} toolbox, and many models for each module, one can obtain almost infinite variety.

1.3 Stochastic versus deterministic models

One of the major current questions in the field of Ca^{2+} modelling is whether to use a stochastic or a deterministic model, and this is a question where the "reality" of the cell's behaviour is of less use than one might think.

High resolution measurements of Ca^{2+} concentration have shown that, in many cell types (most likely all relevant cell types), at low agonist concentrations Ca^{2+} release occurs as a series of small, punctate releases, either from a single IPR (a Ca^{2+} blip), a group of IPR (a puff) or a group of RyR (a spark) [12, 15, 17, 22, 60, 94, 135, 154]. These releases occur stochastically, due to the stochastic opening and closing of the IPR or RyR. If release from one cluster of IPR is large enough, Ca^{2+} can spread to neighbouring clusters of IPR, initiating puffs there, and all the puffs can combine into a global wave [120, 152].

One can now imagine a stochastic scenario for the generation of periodic Ca^{2+} waves. Every so often, just by random chance, one cluster will fire strongly enough to initiate such a global wave. Once the Ca^{2+} concentration returns to baseline after the wave, there will be a random waiting time before the next cluster initiates the next wave, and thus the waiting time between waves, i.e., the wave period, is set by the waiting time between cluster firings, not by any deterministic limit cycle in the dynamics of the cluster.

Such a stochastic mechanism is relatively easily identified experimentally. A purely stochastic wave activation process will result in the wave initiation times being distributed in a Poisson distribution, in which the mean is equal to the standard deviation. Thus, if a plot of the mean wave period versus the standard deviation (for a variety of waves of different periods, found, for example, by using different agonist concentrations) sits close to the line $y = x + b$, for some $b > 0$, this is a clear indication that the waves are being initiated by a Poisson process, with a refractory period (presumably set by some other deterministic process) of b. Note, of course, that if the waves are generated by a purely deterministic process, the standard deviation of the period (for each fixed agonist concentration) is zero.

When one measures the ratio of the mean to the standard deviation (i.e., the *coefficient of variation*, or CV) of the distribution of wave periods, in many cell types the CV turns out to be close to 1. Even for oscillations like those shown in

Fig. 1A, which look to the naked eye as if they are generated by a deterministic process, more detailed studies show that, for a range of IP_3 concentrations, the CV is close to 1 (unpublished results), and thus these oscillations are initiated by a Poisson process. Similar results are found in other cell types [120, 137]. Hence, the weight of evidence suggests that most, and probably all, Ca^{2+} waves are generated by a stochastic, rather than a deterministic process.

However, although this might be the case in reality, the implications for modelling are not clear. It might be tempting to discard all deterministic models as being "wrong", but this would be a poor solution to a difficult question. As is already well established, all models are "wrong", but many remain useful. In fact, deterministic models, despite their lack of stochastic reality, do seem to abstract and describe mechanisms that are crucial for oscillations. Deterministic models have been used in a variety of cell types to make predictions about cell behaviour, and these predictions have been confirmed experimentally, leading, for example, to greatly increased understanding of the interplay between RyR and IPR in airway smooth muscle [149], or the role of Ca^{2+} influx [129].

Recently, a consensus has begun emerging in the Ca^{2+} modelling community that both stochastic and deterministic models are valuable, and that both are needed for a complete understanding of how Ca^{2+} oscillations are generated and controlled. Both are, in essence, putting a face on the actual underlying mechanisms — pumping of Ca^{2+} into the ER, depletion of the ER, Ca^{2+} fluxes through IPR and RyR, and so on — and although the faces differ in detail, the machinery behind them remains similar in many respects. Thus a deterministic model, although ignoring the details of stochastic wave initiation, can remain a highly useful predictive tool, while stochastic models can, in their turn, provide a more solid understanding of exactly how and when each Ca^{2+} spike occurs.

So, with the caveat that deterministic models of Ca^{2+} oscillations and waves must be approached with care, and one should never have too much faith in their immediate physical reality, in the remainder of this article will shall restrict our attention to just such models.

1.4   Excitability

One of the most important features of Ca^{2+} dynamics is the property of Ca^{2+} excitability [74, 88], where a small amount of Ca^{2+} release initiates the release of a larger amount of Ca^{2+}, in a positive feedback process. When first discovered in skeletal muscle this property was called Ca^{2+}-induced Ca^{2+} release, or CICR [41].

CICR can arise in two different ways. Firstly, it can arise through modulation by Ca^{2+} of the IPR or RyR open probability; for example, the open probability curve of the IPR is bell-shaped, increasing at low Ca^{2+} concentrations, and decreasing at high Ca^{2+} concentrations. Thus, at low Ca^{2+} concentrations, an increase in Ca^{2+} concentration leads to an increase in the open probability of the IPR, and hence positive feedback. The details differ between IPR subtypes, but the qualitative behaviour is similar ([50] shows a wide selection of different steady-state curves from various cell types and IPR subtypes, all showing the same fundamental bell shape). At low Ca^{2+} concentrations RyR exhibit a similar behaviour, in that an increase in Ca^{2+} concentration leads to a greater open probability of the RyR and thus CICR. At high Ca^{2+} concentrations the steady-state properties of the RyR are less clear, and there remains controversy over whether the RyR closes again at physiological Ca^{2+} concentrations, and what role such closure might play in excitation-contraction coupling [16, 47, 46, 58, 153].

The second way that CICR can arise is through a dynamic process, in which the activation of the IPR by Ca^{2+} is faster than its inactivation by Ca^{2+} leading to an initial large increase in Ca^{2+} release followed by a slower decline to a lower steady value [36, 48, 50, 66]. In this case, the CICR is a result of the differing time scales of IPR activation and inactivation. If CICR arises from this dynamic process, then it is largely independent of the shape of the steady-state open probability curve. In reality, IPR have both a bell-shaped steady-state curve as well as a time separation between Ca^{2+}-induced activation and Ca^{2+}-induced inactivation. It is thus not necessarily obvious which of these mechanisms underlies any particular experimental observation of CICR; most models, either deterministic or stochastic, incorporate aspects of both mechanisms [2, 33, 43, 120, 126, 137].

As a result of CICR, Ca^{2+} release through IPR and RyR is an autocatalytic, or positive feedback, process, similar in many aspects to the excitability seen in the membrane potential of a neuron [65]. In neurons, the positive feedback occurs via voltage-dependence of the Na^+ channel, which causes a fast depolarisation of the cell. (Excitability of the Na^+ channel arises from the fast activation and slow inactivation of the channel by the membrane potential [73].) Thus, standard excitable models, such as the FitzHugh-Nagumo model, have often been used in models of Ca^{2+} oscillations and waves [21, 127, 141, 142].

However, despite the similarities between the systems, models of Ca^{2+} dynamics differ in important ways from models of other excitable systems. We shall explore some of these differences in this review.

2   ODE models

If a cell is assumed to be well mixed, a typical equation for the Ca^{2+} concentration expresses simply the conservation of Ca^{2+}.

A simple example is shown in Fig. 4. There, the shaded area is the endoplasmic reticulum (ER), and there are five fluxes into or out of the cytoplasm. Two of those fluxes, J_{in} (a generic influx of Ca^{2+}, possibly through store-operated channels, agonist-operated channels, or voltage-dependent Ca^{2+} channels) and J_{pm} (the flux through the plasma membrane ATPase pumps), are across the plasma membrane, while the other three, J_{RyR} (the flux through RyR), J_{IPR} (the flux through IPR)

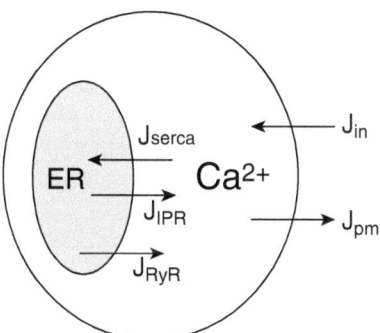

Fig. 4 Schematic diagram of a simple spatially homogeneous Ca^{2+} dynamics model, with five fluxes. In this model, Ca^{2+} is pumped into the ER from the cytoplasm by SERCA pumps (J_{serca}), is pumped out of the cell by plasma membrane ATPase pumps (J_{pm}), enters the cytoplasm from the outside through some unspecified entry pathway (J_{in}), and enters the cytoplasm from the ER through two channels, the IPR and the RyR

and J_{serca} (the flux through the SERCA pumps) are across the ER membrane. (For simplicity we ignore Ca^{2+} buffering for now. This is dealt with in detail in the next section.)

If we let c and c_e denote the Ca^{2+} concentration in the cytoplasm and ER, respectively, with respective volumes V and V_e, then conservation of Ca^{2+} gives

$$\frac{d}{dt}(cV) = J_{in} - J_{pm} + J_{IPR} + J_{RyR} - J_{serca}, \tag{2.1}$$

$$\frac{d}{dt}(c_e V_e) = -J_{IPR} - J_{RyR} + J_{serca}, \tag{2.2}$$

where each J is in units of moles per second.

As long as the volumes of the cytoplasm and ER remain constant, we can divide out the volumes to get

$$\frac{dc}{dt} = \frac{1}{V}(J_{in} - J_{pm} + J_{IPR} + J_{RyR} - J_{serca}), \tag{2.3}$$

$$\frac{dc_e}{dt} = -\frac{1}{V_e}(-J_{IPR} - J_{RyR} + J_{serca}). \tag{2.4}$$

In simple models like this it is usual to rescale all the fluxes, so that they have units of moles per time per cytoplasmic volume. Thus, we define, for example, a new $\tilde{J}_{in} = J_{in}/V$, and rewrite both equations in these new units.

If we do this, and then (for notational convenience) drop the tildes, we get

$$\frac{dc}{dt} = J_{IPR} + J_{RyR} - J_{serca} + J_{in} - J_{pm}, \tag{2.5}$$

$$\frac{dc_e}{dt} = \gamma(-J_{\text{IPR}} - J_{\text{RyR}} + J_{\text{serca}}), \qquad (2.6)$$

where $\gamma = \frac{V}{V_e}$. The factor of γ appears since the flux of x moles from the cytoplasm to the ER causes a different change in concentration in each compartment, due to their different volumes. Each J is in units of moles per cytoplasmic volume per second.

Now one selects whichever model one wishes for each of the individual fluxes to complete the model construction. In general, each of these fluxes will involve other dynamic variables, which increases the total number of differential equations. Simpler models will have only two equations, more complex models typically have as many as eight, or even more.

We emphasise that, although this simple model omits a vast amount of the known complexity in Ca^{2+} signalling (such as microdomains, the influence of the mitochondria, and direct effects of the membrane potential), it is still (as we shall see) a useful tool for the study of the mechanisms underlying Ca^{2+} oscillations, in some conditions.

2.1 Calcium buffering

Calcium is heavily buffered in all cells, with at least 99% (and often more) of the available Ca^{2+} bound to large Ca^{2+}-binding proteins. For example, calsequestrin and calreticulin are major Ca^{2+} buffers in the ER and SR, while in the cytoplasm Ca^{2+} is bound to calbindin, calretinin and parvalbumin, among many others. Calcium pumps and exchangers and the plasma membrane itself are also major Ca^{2+} buffers. In essence, a free Ca^{2+} ion in solution in the cytoplasm cannot do much, or go far, before it is bound to something.

The basic chemical reaction for Ca^{2+} buffering can be represented by the reaction

$$P + Ca^{2+} \underset{k_-}{\overset{k_+}{\rightleftarrows}} B, \qquad (2.7)$$

where P is the buffering protein and B is buffered Ca^{2+}. Letting b denote the concentration of buffer with Ca^{2+} bound, and c the concentration of free Ca^{2+}, a simple model of Ca^{2+} buffering is

$$\frac{dc}{dt} = f(c) + k_- b - k_+ c(b_t - b), \qquad (2.8)$$

$$\frac{db}{dt} = -k_- b + k_+ c(b_t - b), \qquad (2.9)$$

where k_- is the rate constant for Ca^{2+} release from the buffer, k_+ is the rate constant for Ca^{2+} uptake by the buffer, b_t is the total buffer concentration and $f(c)$ denotes all the other reactions involving free Ca^{2+} (release from the IP_3 receptors, reuptake by pumps, etc.). Note that, from conservation of buffer molecules, $[P] + b = b_t$.

If the buffer has fast kinetics, its effect on the intracellular Ca^{2+} dynamics can be analysed simply [147]. If k_- and k_+c_0, where c_0 is some natural scale for the Ca^{2+} concentration (often around 1 μM), are large compared to the time constant of Ca^{2+} reaction (for example, the speed of release through the IPR or uptake by SERCA pumps), we take b to be in the quasi-steady state

$$k_-b - k_+c(b_t - b) = 0, \tag{2.10}$$

and so

$$b = \frac{b_t c}{K + c}, \tag{2.11}$$

where $K = k_-/k_+$. Adding (2.8) and (2.9), we find the "slow" equation

$$\frac{d}{dt}(c + b) = f(c), \tag{2.12}$$

which, after using (2.11) to eliminate b, becomes

$$\frac{dc}{dt} = \frac{f(c)}{1 + \theta(c)}, \tag{2.13}$$

where

$$\theta(c) = \frac{b_t K}{(K + c)^2}. \tag{2.14}$$

Note that we assume that b_t is a constant. Hence, fast Ca^{2+} buffering simply adds a Ca^{2+}-dependent scaling factor to all the fluxes.

If the buffer is not only fast, but also of low affinity, so that $K \gg c$, it follows that

$$\theta \approx \frac{b_t}{K}, \tag{2.15}$$

a constant. Such a constant, multiplying all the fluxes in the model, can be simply incorporated into the other rate constants, and ignored henceforth, with the proviso that all fluxes must be interpreted as effective fluxes, i.e., that portion of the actual flux that contributes to a change in free $[Ca^{2+}]$. Hence although it might appear at first glance that equation (2.5) ignores Ca^{2+} buffering, that is not the case. Rather, it is just assuming that Ca^{2+} buffering is fast and linear, and thus that all fluxes are effective fluxes.

There have been a number of studies of the effects of nonlinear buffers on the dynamics of Ca^{2+} oscillations (for example, see [52] or [42]), but these results are beyond the scope of this review. In general, the qualitative effects of nonlinear buffers are small, except in certain narrow parameter regimes. In this review we shall mostly just assume that buffering is fast and linear, and thus does not appear explicitly. An asymptotic analysis of Ca^{2+} buffering was performed by [121]; other important theoretical papers on Ca^{2+} buffering are [98, 100, 122, 123, 128, 141].

2.2 Modelling the calcium fluxes

In order to construct a specific realisation of (2.5) and (2.6), we need first to decide how to model each of the calcium fluxes in those equations. Since there is an enormous range of possible choices, we shall focus only on a few selected models.

2.2.1 IPR fluxes

Probably the most important, and the most difficult to model, fluxes are those through the IPR and RyR. IPR models have had a long and complicated history, starting from the earliest models of [33] and [55], through to the most recent models based on single-channel data [18, 118, 116]. Earlier models are reviewed in [126].

All these models share one crucial feature – that the steady-state open probability of the IPR is a bell-shaped function of $[Ca^{2+}]$ (Fig. 5), as has been shown

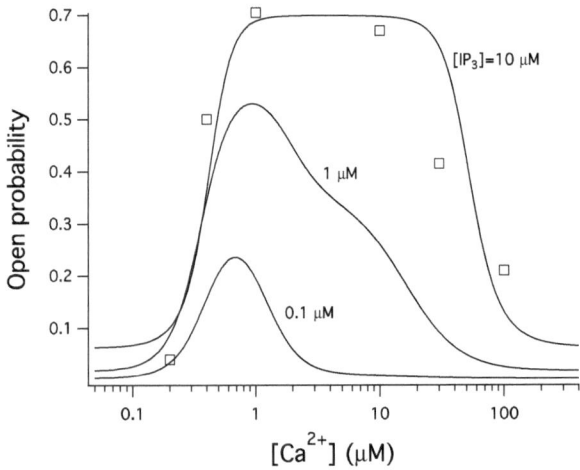

Fig. 5 Open probability (P_o) of the IPR as a function of Ca^{2+} is bell-shaped, increasing at lower $[Ca^{2+}]$ and decreasing at higher $[Ca^{2+}]$. Open squares are data from type I IPR, measured at 10 μM $[IP_3]$ [148], and the smooth curves are from the model of [18]

experimentally many times [50]. However, there are many different ways of attaining such a steady-state curve, and many different ways of modelling the dynamic features of the IPR.

In earlier models [33, 2, 125], the most important basic dynamic property of IP$_3$ receptors is that they respond in a time-dependent manner to step changes of Ca^{2+} or IP$_3$. Thus, in response to a step increase of IP$_3$ or Ca^{2+} the receptor open probability first increases to a peak and then declines to a lower plateau. This decline is called *adaptation* of the receptor, since the open probability adapts to a maintained Ca^{2+} or IP$_3$ concentration. If a further step is applied on top of the first, the receptor responds with another peak, followed by a decline to a plateau. In this way the IPR responds to *changes* in [Ca^{2+}] or [IP$_3$], rather than to absolute concentrations.

One popular model is one of the earliest, due to De Young and Keizer [33]. In this model, it is assumed that the IP$_3$ receptor consists of three equivalent and independent subunits, all of which must be in a conducting state for there to be Ca^{2+} flux. Each subunit has an IP$_3$ binding site, an activating Ca^{2+} binding site, and an inactivating Ca^{2+} binding site, each of which can be either occupied or unoccupied, and thus each subunit can be in one of eight states.

Simplification by Li and Rinzel [86] of this eight-state model led to the model

$$P_o = \left(\frac{pcr}{(p + K_1)(c + K_5)} \right)^3, \tag{2.16}$$

$$\tau_r(c,p)\frac{dr}{dt} = r_\infty(c,p) - r, \tag{2.17}$$

where P_o is the open probability, p is [IP$_3$], K_1 and K_5 are constants, and r is the fraction of receptors that have not been inactivated. The functions τ_r and r_∞ are given in detail in [86]. Writing the model in this form emphasises the mathematical similarities with the model of the Na$^+$ channel in the Hodgkin-Huxley model [65], thus highlighting their common feature of excitability.

A similar model, that appeared at the same time as the De Young and Keizer model, is due to Atri et al. [2] and takes a slightly simpler form. In the Atri model, the open probability of the IPR is assumed to take the form

$$P_o = k_f \left(\mu_0 + \frac{\mu_1 p}{k_\mu + p} \right) \left(b + \frac{(1-b)c}{k_1 + c} \right) r, \tag{2.18}$$

$$\tau\frac{dr}{dt} = \frac{k_2^2}{k_2^2 + c^2} - r. \tag{2.19}$$

Thus, P_o is an increasing function of the IP$_3$ concentration, and, over fast time scales, an increasing function of c also. However, on the time scale set by τ, r acts as a Ca^{2+}-dependent inactivation variable, and causes Ca^{2+}-dependent and time-dependent inactivation of the receptor. (As in the Li-Rinzel model, r denotes the fraction of receptors that have not been inactivated by Ca^{2+}). Overall, this model gives a bell-shaped steady-state open probability curve, as seen experimentally, but has no satisfactory biophysical basis for the various terms.

Mathematical studies of Ca^{2+} dynamics have tended to use early IPR models, such as the ones described above. However, the most recent data have shown that the details of these early IPR models are not correct. We now know that the IPR exists in two (or possibly more) "modes" [18, 67, 93, 116, 118]. In one mode (sometimes called the Park mode) the receptor is mostly closed, while in the other mode (the Drive mode) the receptor is mostly open. Transitions between the two modes are controlled by $[Ca^{2+}]$, $[IP_3]$ and $[ATP]$, among other things, but transitions within each mode are independent of these ligands. Such modal behaviour cannot be reproduced by most early models, which have the incorrect Markov structure. In addition, the early models do not usually give the open-time and closed-time distributions (to choose two statistics in particular) that have been observed in the most recent single-channel data from nuclear patch clamp studies.

Nevertheless, although the details of the early models are incorrect, the fundamental IPR properties remain the same. For example, [18] has shown that fast Ca^{2+}-induced activation followed by slow Ca^{2+}-induced inactivation remain as crucial ingredients in these recent modal models.

For this reason, we shall focus here on mathematical studies of Ca^{2+} models based on older IPR models. When the newer generation of IPR models come to be incorporated into whole-cell models, the mathematical techniques (and dangers thereof) will remain the same.

2.2.2 RyR fluxes

The selection of RyR models is similarly complex. Some models [51] are based on simple and heuristic CICR, and fit data well, while a variety of other models, mostly designed for use in cardiac cell models [57, 56, 132, 151], incorporate multiple receptor states and stochastic behaviour. Because the literature on cardiac cells, skeletal muscle, RyR models and excitation-contraction coupling is so vast, we cannot even begin to do it justice in this review. Thus, we shall take the opposite approach and simply not discuss these areas at all (except in some restricted cases). The reviews of [8, 46] give excellent entries to the field, as do [7, 17, 16, 56, 115, 132, 145, 151].

2.2.3 Calcium pumps

Experimental data indicate that SERCA pumps transfer two Ca^{2+} ions across the ER/SR membrane per cycle [14, 89, 97, 138]. Thus, the most common way to model the Ca^{2+} flux, J_{serca}, due to SERCA pumps, is to use a simple Hill equation, with Hill coefficient of two. Thus,

$$J_{\text{serca}} = \frac{V_m c^2}{K_m^2 + c^2}.$$

The parameter K_m we know to be approximately $0.27\ \mu M$, while V_m, which depends on the density of SERCA pumps, can vary substantially depending on the cell type.

It is worth noting that this equation for J_{serca} contains within it a host of simplifications. More detailed models of SERCA pumps [63, 77, 89, 138] involve multiple states, with the pump protein moving between these states to pick up Ca^{2+} ions on one side of the ER membrane and release them on the other. More accurate models of SERCA pumps would take these states into account, as well as keeping track of all the Ca^{2+} bound to the pump protein. ([73] gives an introductory discussion of a range of SERCA models, ranging from the simplest, to more complex versions.) Although such detailed models appear to cause little change in dynamic behaviour [63] one should keep in mind that the simplifications used to obtain J_{serca} (for example, quasi-steady-state approximations) are of the exact same type as those used to simplify Ca^{2+} models, as discussed in this review, and come with all the same caveats and potential for complications.

2.2.4 Calcium influx

Over recent years it has become clear that the influx of Ca^{2+} into the cell from outside is no simple matter [112, 117, 124, 131]. It is controlled by a variety of proteins that are themselves controlled by a variety of factors such as arachidonic acid, or the concentration of Ca^{2+} in the ER. For some of these influx pathways geometrical factors, such as the close apposition of the ER and the plasma membrane, play a significant role.

However, for the purposes of the discussion here, we can divide all Ca^{2+} influx pathways into three major types.

1. Voltage-dependent Ca^{2+} channels, or VDCCs [19]. These open in response to depolarisation of the cell membrane, and play a vital role in excitable cells such as skeletal and cardiac muscle, in some smooth muscle cells, in neuroendocrine cells, and in a variety of neuronal cell types.
2. Receptor-operated channels, or ROCs [70]. Some Ca^{2+} influx pathways open in response to agonist stimulation, often via the production of arachidonic acid. Thus, Ca^{2+} influx is usually modelled as an increasing function of agonist concentration. The exact mechanisms of this dependency are, in general, unknown, so detailed models of ROCs are not realistically possible at this stage.
3. Store-operated channels, or SOCs [103]. Severe depletion of the ER causes the opening of Ca^{2+} channels in the cell membrane, via a process involving ORAI and STIM molecules. This is an important influx pathway under conditions of high prolonged agonist concentration, but will play little role in our analysis here.

2.3 Model classification

2.3.1 Open cell/Closed cell models

One common experimental technique is to remove Ca^{2+} from outside the cell, and observe how this affects the intracellular Ca^{2+} oscillations. In many cases the oscillations continue for a considerable time before finally running down (due to the progressive loss of Ca^{2+} from the cell), while in other cases the oscillations are terminated immediately. This has motivated the detailed study of the effects of Ca^{2+} entry on oscillatory properties.

To study the effects of Ca^{2+} entry, models are generally constructed in two different classes.

- Open cell models are those in which Ca^{2+} is allowed to enter and leave the cell freely across the plasma membrane. Thus, such models include Ca^{2+} influx pathways and plasma membrane Ca^{2+} pumps, and the total amount of Ca^{2+} in the cell is not conserved.
- Closed cell models are those in which all Ca^{2+} transport across the plasma membrane, both inward and outward, is blocked. Note that a closed cell model does not correspond exactly to the experimental situation of low external Ca^{2+} concentration, but will approximate the situation at the beginning of the experiment. It is possible experimentally to block the plasma membrane Ca^{2+} pumps also, using high concentrations of ions such as lanthanum, but these are more difficult experiments to perform and more difficult to interpret, due to the varied effects of lanthanum.

2.3.2 Class I/Class II models

The second way in which Ca^{2+} oscillation models are typically classified is with respect to the behaviour of IP_3. In some cell types, Ca^{2+} oscillations occur when IP_3 concentration is constant, and such oscillations are believed to be caused by the intrinsic dynamics (i.e., the fast activation and slower inactivation by Ca^{2+}) of the IPR [130]. Models of such oscillations are called Class I models. In other cell types, Ca^{2+} oscillations are necessarily accompanied by IP_3 oscillations, and if those IP_3 oscillations are blocked, so are the Ca^{2+} oscillations. In such cells, the feedback loops illustrated in Fig. 3 are an integral part of the oscillation mechanism. Such models are called Class II models. Models which partake both of Class I and Class II properties are called *hybrid* models [35]. Although, realistically, every cell type will have both Class I and Class II mechanisms to differing degrees, and thus should be modelled by a hybrid model, it is useful to make this distinction, and to study the behaviour of pure Class I and II models.

It is also important to note that Ca^{2+} oscillations can also be generated by the entry and exit of Ca^{2+} from the cell. Such oscillations cease immediately upon removal of extracellular Ca^{2+}, and thus require an open-cell model. However, models of this type are neither Class I nor Class II. A simple example of this type of model is discussed in Section 5.3.

2.4   A simple example: the combined model

All these concepts, and the various types of models, can be simply illustrated by a single set of equations [35]. For convenience, we shall call this model the *combined* model, as it combines both Class I and Class II mechanisms in such a way that it is simple to switch from one class of model to the other.

As usual, we let c and c_e denote, respectively, the concentrations of Ca^{2+} in the cytoplasm and the ER, we let p denote the IP_3 concentration, and we let r denote the fraction of IPR that have not been inactivated by Ca^{2+} (as in the Atri model described above).

$$\frac{dc}{dt} = J_{IPR} - J_{serca} + \delta(J_{influx} - J_{pm}), \tag{2.20}$$

$$\frac{dc_e}{dt} = \gamma(-J_{IPR} + J_{serca}), \tag{2.21}$$

$$\frac{dp}{dt} = \nu\left(1 - \frac{\alpha k_4}{c + k_4}\right) - \beta p, \tag{2.22}$$

$$\frac{dr}{dt} = \frac{1}{\tau}\left(\frac{k_2^2}{k_2^2 + c^2} - r\right), \tag{2.23}$$

where

$$J_{IPR} = \left[k_{flux}\left(\mu_0 + \frac{\mu_1 p}{k_\mu + p}\right)\left(b + \frac{V_1 c}{k_1 + c}\right)r\right](c_e - c), \tag{2.24}$$

$$J_{serca} = \frac{V_e c}{K_e + c}, \tag{2.25}$$

$$J_{pm} = \frac{V_p c^2}{k_p^2 + c^2}, \tag{2.26}$$

$$J_{influx} = \alpha_1 + \alpha_2\frac{\nu}{\beta}. \tag{2.27}$$

We note a number of things about this model.

- It uses the Atri model of the IPR [2], and the IPR flux is multiplied by the term $c_e - c$, so that it depends on the Ca^{2+} concentration gradient between the ER and the cytoplasm. It could, just as easily, have used one of the other IPR models in the literature, and the results would, by and large, be qualitatively similar; our choice of the Atri model is purely for simplicity.

- The SERCA pumps are modelled by a Hill function, with Hill coefficient 1. This ignores cooperativity in the SERCA pumps, and thus is not the most accurate assumption that can be made, but it simplifies the analysis somewhat, and has little effect on the results we present here.
- The IP$_3$ concentration, p, obeys its own differential equation, where the production of p can be Ca^{2+}-dependent, as long as $\alpha \neq 0$. However, if $\alpha = 0$, the equation for p essentially decouples. Hence, $\alpha = 0$ corresponds to a Class I model.
- In the limit as $\tau \to 0$, r becomes an algebraic function of c. Thus, the case $\tau \to 0$ and $\alpha \neq 0$ corresponds to a Class II model, in which any oscillations are governed by the interactions between c and p, not by the dynamics of the IPR.
- The parameter δ is introduced so that the rate of Ca^{2+} transport across the plasma membrane can be easily controlled. In many cell types δ is small compared to the time scales of Ca^{2+} transport and release through the IPR and SERCA pumps.
- The parameter ν corresponds to the maximal rate of IP$_3$ production, and is a surrogate for the agonist concentration; as the agonist concentration increases, both the rate of production of IP$_3$ and the rate of Ca^{2+} influx from the outside increases. Thus, in this model, Ca^{2+} influx is via receptor-operated channels. J_{influx} is a linear function of agonist concentration, an expression which has no biophysical basis, but is merely the simplest possible way to make Ca^{2+} influx increase with agonist. As usual with models like this, many of the terms are suggestive of what we believe are the actual mechanisms, but should not be interpreted too literally.

A useful approach, that accentuates the difference between open cell models and closed cell models, is to express the model in terms of a new variable, $c_t = c + c_e/\gamma$, where γ is the ratio of cytoplasmic to ER volume, as defined after eqn. (2.6). Thus, c_t is the total number of moles of Ca^{2+} in the cell, divided by the cytoplasmic volume, and is a measure of the Ca^{2+} load of the cell, i.e., how much Ca^{2+} the cell contains. Using this new variable the first two model equations become

$$\frac{dc}{dt} = J_{\text{IPR}} - J_{\text{serca}} + \delta(J_{\text{influx}} - J_{\text{pm}}), \tag{2.28}$$

$$\frac{dc_t}{dt} = \delta(J_{\text{influx}} - J_{\text{pm}}). \tag{2.29}$$

It is now clear that, as δ becomes smaller, c_t becomes a slower variable than c, and in the limit of $\delta = 0$ we obtain a closed cell model.

Hence, by varying α, δ and τ we can use this single set of equations to illustrate both open and closed cell models, as well as Class I and Class II models.

3 Bifurcation structure of ODE models

A natural first step towards understanding the dynamics of models such as the combined model (i.e., equations (2.20)–(2.23)) is to construct a bifurcation diagram; this allows us to locate parameter regimes in which behaviour of interest, such as calcium oscillations, can occur. For most models, there are many possible choices of bifurcation parameter, but it is common to choose as the main bifurcation parameter a quantity that corresponds to something that is relatively easy to manipulate experimentally. Doing so makes it easier to compare model output to experimental results, and thus to validate the model or use model predictions to inform experiments. For instance, in the combined model, we can choose v as the primary bifurcation parameter; v corresponds to the maximal rate of IP$_3$ production, which is relatively easy to modify in an experiment since it is an increasing function of the level of agonist applied to the cell.

Fig 6A shows a partial bifurcation diagram for the combined model, for the choice $\alpha = 1$, $\tau = 2$ (i.e., a hybrid version of the model) and other parameters as specified in Table 1 in the Appendix. Time series and phase portraits for two choices of v are shown in the other panels. This bifurcation diagram is typical of many models of intracellular calcium dynamics, in the sense that we see no oscillations of [Ca^{2+}] for sufficiently small or sufficiently large v, but there is a region of intermediate parameter values (between the points labelled HB$_1$ and HB$_2$) in which there is a variety of different types of oscillation. This is what is seen experimentally; at low agonist concentrations, there is not enough IP$_3$ to open the IPR, while at high agonist concentration, there is so much IP$_3$ in the cell, and such a high resting [Ca^{2+}], that the IPR is again kept shut (remember that the steady-state open probability curve of the IPR is bell-shaped, and so the IPR is closed at both low and high [Ca^{2+}]).

One feature common to both the time series shown is that there are time intervals in which there is very rapid increase or decrease of calcium concentration interspersed with intervals of much slower change. These are typical solutions for models with more than one time scale; in the case of the combined model, this results in part from the choice $\delta = 0.01$, which causes the variable c_t to evolve much more slowly than the variable c, at least for certain choices of the bifurcation parameter, v, and in certain regions of the phase space. Methods for the analysis of mathematical models with multiple time scales are well developed in general, although only recently applied in a systematic way to models of calcium dynamics; these methods are discussed further in section 3.1. For now, we note only that the oscillation shown in panels B and C of Fig 6 is a *relaxation oscillation*, while that shown in panels D and E is a *mixed mode oscillation*, and has a number of small, subthreshold oscillations occurring between each pair of large spikes in calcium concentration. Note that the subthreshold oscillations in panel D are of very small amplitude, and are essentially invisible on the scale of the main panel. However, the presence of these tiny oscillations can have a marked effect on the observed dynamics, as will be discussed further in section 3.2. At values of v close to HB$_2$, it

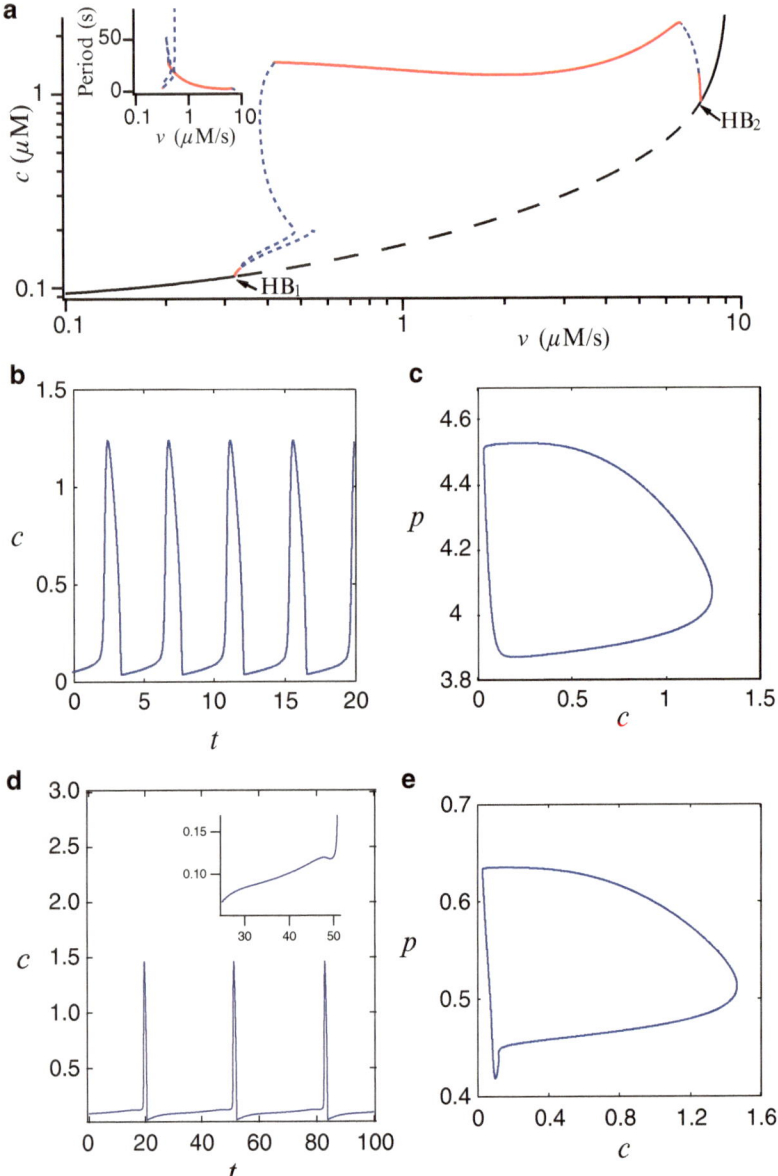

Fig. 6 Partial bifurcation diagram and some time series and phase portraits for equations (2.20)–(2.27), with $\alpha = 1$, $\tau = 2$ and other parameter values as in Table 1. Panel A shows the bifurcation diagram, plotting the cytosolic calcium concentration, c, versus the maximal rate of IP$_3$ formation, v. The black curve shows the position of the steady state solution (dashed curve when it is unstable, solid curve when it is stable). The red and blue curves indicate the maximum amplitudes of stable and unstable periodic orbits, resp. Hopf bifurcations are labelled HB. The inset shows the period of the branches of periodic orbit plotted in the main panel. Panels B and D show time series for c for the attracting periodic solutions that occur at $v = 2.0$ and $v = 0.4$, resp., with the insert to panel D showing an enlargement of part of the time series. Panels C and E show the same solutions as in B and D, resp., projected onto the c-p plane

is also possible to see attracting quasiperiodic oscillations, a feature quite commonly seen in calcium models. Further detail about the bifurcations associated with this and related models is contained in [35, 61] and [62].

Fig. 7A shows an analogous bifurcation diagram for the Class II version of the same model, i.e., with $\alpha = 1$ and in the limit $\tau \to 0$. As can be seen, the range of ν values for which oscillations occur is much smaller than for the hybrid model, and evidence in the time series for the existence of different time scales

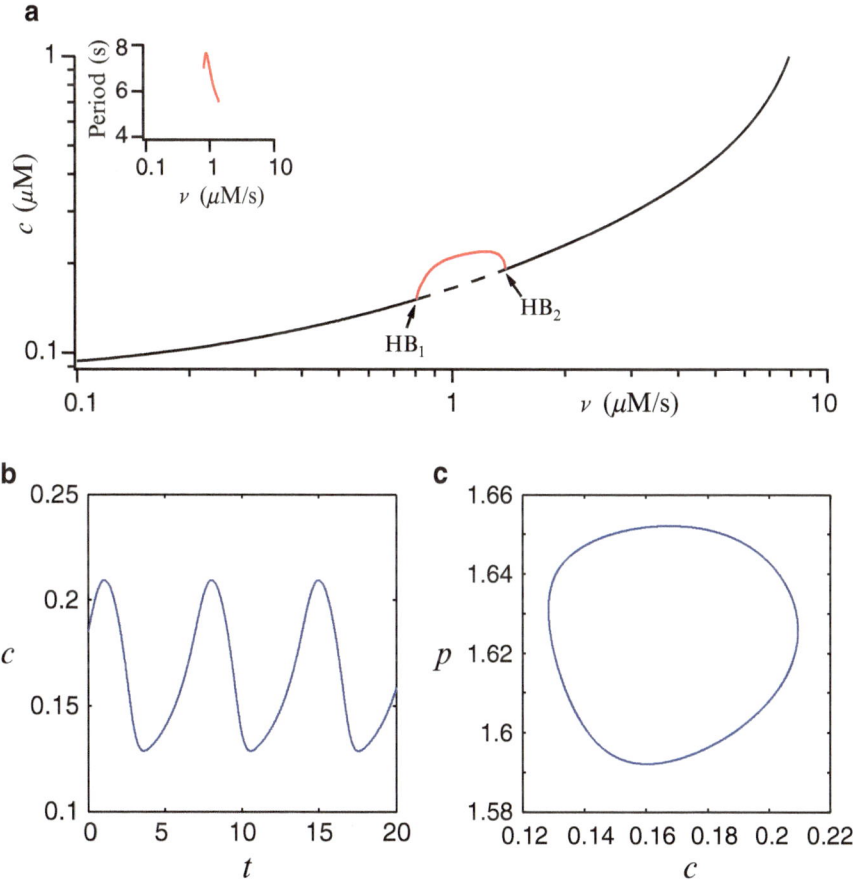

Fig. 7 Partial bifurcation diagram and a corresponding time series and phase portrait for the combined model (equations (2.20)–(2.23)), with $\alpha = 1$ and in the limit $\tau \to 0$, and with other parameter values as in Table 1. Panel A shows the bifurcation diagram, plotting the cytosolic calcium concentration, c, versus the maximal rate of IP$_3$ formation, ν. The inset shows the period of the branch of periodic orbits. Panel B shows a time series for c, for the attracting periodic solution that occurs at $\nu = 1.0$, and panel C shows the same solution projected onto the c-p plane. Line styles and labels are as in Fig 6

is less pronounced. Similar time series (although with differing amplitudes of the oscillations) occur for all values of ν for which there are oscillations.

Note that Fig 6A is an incomplete bifurcation diagram for the associated model; just the primary branches of periodic orbits are shown, and there are a number of bifurcations along these branches that have not been identified here (e.g., at each place where the stability of a periodic orbit branch changes). While a detailed knowledge of the bifurcation structure of a model may be of interest from a mathematical point of view, the details are frequently irrelevant from the point of view of understanding the underlying physiology. Data from typical experiments might consist of noisy time series of $[Ca^{2+}]$ (or possibly both $[Ca^{2+}]$ and $[IP_3]$), from which an approximate amplitude and frequency of the oscillation can be extracted, but these may not be able to be directly compared with predictions from the model, due to the large number of unknown parameters in the model. Furthermore, unstable solutions will not be directly observed, and experimental time series may not have enough precision or length to resolve other details, such as subthreshold oscillations. We note, however, that an understanding of the mathematical details of model dynamics, including unstable solutions, sometimes provides crucial insight into physiological mechanisms that may underlie experimental observations; an example of such a case is discussed in section 3.2.

3.1   Fast-slow reductions

ODE models of calcium dynamics frequently exhibit behaviour indicative of the presence of different time scales in the problem, as discussed above, and a variety of techniques that exploit the time scale separation may be helpful in the analysis of these models. A first step in the process is identification of the time scales present in the model. Sometimes, an understanding of the physiology underlying the model assists this process. For instance, in many situations, the variation of the total calcium (c_t) in a cell is known to be slow relative to variations in cytoplasmic or ER calcium concentrations (this was discussed above in the context of the combined model) and c_t can then be designated as a slow variable. There are good physiological reasons for this; as discussed in Section 1.1, cells expend a great deal of energy keeping cytoplasmic $[Ca^{2+}]$ low, against a very large $[Ca^{2+}]$ gradient. It is thus desirable for cells to restrict severely the ability for Ca^{2+} to cross the cell membrane. Hence, background Ca^{2+} influx into cells tends to be very slow, to be matched by an equally slow background Ca^{2+} removal from the cell.

However, beyond the designation of total calcium as a slow variable, the situation can be quite complicated: there may be more than one slow variable or more than two time scales, the relative speed of evolution of the variables may change within the phase space, and intuition based on physiological considerations may be misleading.

From a mathematical point of view, an approach that is frequently helpful is to non-dimensionalise the model equations, then group variables according to their relative speed of evolution in the non-dimensional version of the model. For instance, as discussed in [62], a non-dimensional form of equations (2.22), (2.23), (2.28) and (2.29) can be obtained by introducing new dimensionless variables, (C, C_t, P, t_1) with

$$c = Q_c \cdot C, \quad c_t = Q_c \cdot C_t, \quad p = Q_p \cdot P, \quad t = T \cdot t_1,$$

where $Q_c = 1 \ \mu M$ and $Q_p = 10 \ \mu M$ are typical concentration scales for calcium and IP$_3$, resp., and $T = Q_c/(\delta V_P) = 100/24$ s is a typical time scale for the c_t dynamics. (Note that the variable r was already dimensionless in the original model.) This then leads to rescaled evolution equations:

$$\delta \frac{dC}{dt_1} = \bar{J}_{\text{release}} - \bar{J}_{\text{serca}} + \delta(\bar{J}_{\text{in}} - \bar{J}_{\text{pm}}),$$

$$\frac{dC_t}{dt_1} = \bar{J}_{\text{in}} - \bar{J}_{\text{pm}}, \tag{2.30}$$

$$\frac{dr}{dt_1} = \frac{1}{\hat{\tau}} \left(\frac{k_2^2}{k_2^2 + Q_c^2 C^2} - r \right),$$

$$\frac{dP}{dt_1} = \hat{v} \left(1 - \frac{k_4 \alpha}{k_4 + Q_c C} \right) - \hat{\beta} P,$$

with dimensionless parameters

$$\hat{\tau} = \frac{\delta V_p}{Q_c} \tau, \quad \hat{v} = \frac{Q_c}{Q_p} \frac{v}{\delta V_p}, \quad \hat{\beta} = \frac{Q_c}{\delta V_p} \beta, \tag{2.31}$$

and corresponding dimensionless versions of the fluxes, $\bar{J}_{\text{release}}$, $\bar{J}_{\text{serca}}$, $\bar{J}_{\text{pm}}$ and $\bar{J}_{\text{in}}$. With the choice of parameters values given in Table 1, and for v values corresponding to oscillatory solutions, we then find that the speeds of evolution for the variables are $O(10^2)$ for C, $O(1)$ for C_t and P, and order $O(1/\hat{\tau})$ for r. Thus, if $\hat{\tau}$ is $O(1)$, this system has one fast variable and three slow variables, while if $\hat{\tau}$ is $O(\delta)$, there are two fast variables and two slow variables.

A common next step in the analysis of certain classes of model is to remove fast variables using a quasi-steady state (QSS) approximation. The idea is that certain variables may evolve so fast that their evolution equations can be replaced by algebraic equations, thereby reducing the dimension of the model. For instance, in equations (2.30), if $\hat{\tau}$ is small enough (e.g., $O(10^{-3})$ or smaller), then r can be regarded as the fastest variable of the model, and we might assume that $dr/dt_1 \approx 0$ so that

$$r \approx \frac{k_2^2}{k_2^2 + Q_c^2 C^2}.$$

The QSS approximation replaces r in the model by its QSS value, $r_\infty(C)$:

$$r_\infty \equiv \frac{k_2^2}{k_2^2 + Q_c^2 C^2}.$$

The model then reduces to three differential equations, and becomes a Class II version of the model as discussed in section 2.4.

Although appealing from a modelling perspective, use of a QSS approximation can lead to difficulties. As discussed in [157], QSS reduction can remove a Hopf bifurcation from the dynamics or change the position or criticality of a Hopf bifurcation. In such cases, the occurrence and/or nature of oscillations in the reduced model may be significantly different to that for the original model, usually an undesirable outcome. Further work on the effect of QSS reduction is underway, but early results [157] suggest that singular Hopf bifurcations [13] (in which both fast and slow variables are involved in the bifurcation, and which are common in models of biophysical systems) may be relatively unaffected by QSS reduction.

A different reduction technique that has had some success in explaining the dynamics of calcium models involves effectively removing one of the slow variables by treating it as a parameter. This method was pioneered by Rinzel [108] in his classic study of bursting electrical oscillations in pancreatic beta cells, and has since been widely used in the study of a range of oscillating biophysical models. The idea is that characteristics of an attracting solution occurring at a particular value of the genuine bifurcation parameter can be understood by comparing it with the bifurcation diagram obtained by fixing the genuine bifurcation parameter but using the slowest variable as a parameter.

For instance, for the Class II version of the combined model expressed in (c, c_t, p) coordinates, i.e., equations (2.22), (2.28) and (2.29) with $\alpha = 1$ and $r(c) = \frac{k_2^2}{k_2^2 + c^2}$, the variable c_t is slowest so long as δ is sufficiently small. One can then remove the dc_t/dt equation and treat c_t as a constant where it appears elsewhere in the model, then construct a bifurcation diagram using c_t as the bifurcation parameter. For the choice $\nu = 1.0$, this method results in the bifurcation diagram shown in black in Fig 8. After superimposing on this bifurcation diagram the attracting orbit of the full problem, with c_t allowed to vary but with ν still fixed at the same value, and with $\delta = 0.0001$, it can be seen that the orbit (shown in red in Fig 8) moves slowly near the stable branches of the bifurcation diagram, in a direction determined by the true value of dc_t/dt, and makes fast jumps between branches when it reaches the end of a stable section of the bifurcation diagram. In this way, the fast-slow nature of the orbit of the original problem with $\delta = 0.0001$ can be 'understood' in terms of the bifurcation diagram of the system obtained by 'freezing' the slow variable c_t. By varying the value of the fixed (genuine) bifurcation parameter, one can then explain transitions between different types of orbit in the full system.

There are a number of potential difficulties with the use of this 'frozen' system approach. First of all, it presumes that a single globally valid slowest variable can be identified; in reality, variables may have different relative speeds of evolution in

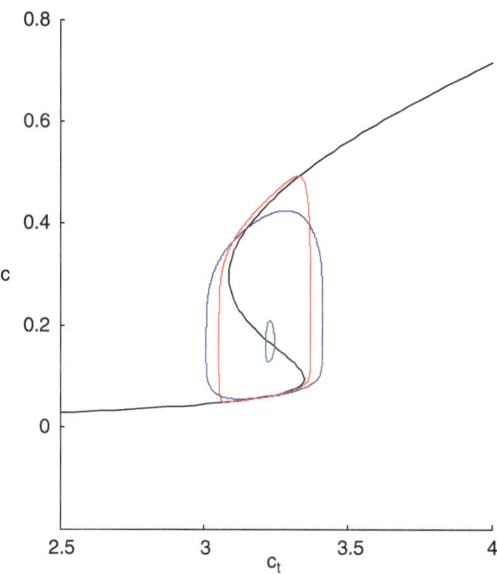

Fig. 8 Bifurcation diagram of the 'frozen' Class II combined model, equations (2.22) and (2.28) with $r(c) = \frac{k_2^2}{k_2^2+c^2}$ and c_t treated as the bifurcation parameter. Parameter values are $v = 1.0$, $\alpha = 1$ and other constants as in Table 1. The black curve indicates steady states of the frozen system; the upper and lower branches are stable, the middle branch is unstable. The red (resp. blue and green) curve shows a solution of the full Class II system for $\delta = 0.0001$ (resp. 0.001 and 0.01)

different parts of the phase space and at different values of the bifurcation parameter. Even if a slowest variable is identifiable, it may not be sufficiently slow for the method to be useful. For instance, Fig 8 shows orbits of the Class II combined model for the choices $\delta = 0.01, 0.001$ and 0.0001. Without a proper time scale analysis, it is not known in advance how small δ needs to be for the 'frozen' system approach to be useful, but it is apparent from Fig 8 that $\delta = 0.01$ is not small enough (the orbit with this value of δ does not follow branches of the frozen bifurcation diagram) and that $\delta = 0.001$ is marginal.

Second, many systems have more than one variable evolving on the slowest time scale; while it is possible to adapt the method to the case of two slow variables, the method rapidly becomes cumbersome. Third, this method may not give accurate information about the regions of transition from fast to slow sections of an orbit, which occur when the distinction between 'fast' and 'slow' variables is lost; these regions are often highly significant for distinguishing between different mechanisms in the dynamics (for instance, the difference between the relaxation oscillations and mixed mode oscillations shown in Fig. 6 occurs precisely at the point where the oscillations change from fast to slow evolution, and these differences are crucial for understanding some phenomena (see section 3.2)). Finally, limited information is provided by the method about the robustness of orbits to changes in the genuine bifurcation parameter.

A more rigorous approach involves the use of geometric singular perturbation theory (GSPT). The idea is to define one or more small parameters in the model. By regarding the model system as a perturbation from a limiting case in which the small parameter(s) tend to zero, it may be possible to extract useful information about the mechanisms underlying complicated dynamics in the original model. For instance, for system (2.30) in the case that $\hat{\tau}$ is $O(\delta)$, we can introduce a small singular perturbation parameter ϵ, and rewrite the model as

$$\epsilon \frac{dC}{dt_1} = \bar{J}_{\text{release}} - \bar{J}_{\text{serca}} + \delta(\bar{J}_{\text{in}} - \bar{J}_{\text{pm}})$$

$$\frac{dC_t}{dt_1} = \bar{J}_{\text{in}} - \bar{J}_{\text{pm}} \tag{2.32}$$

$$\frac{dr}{dt_1} = \frac{1}{\hat{\tau}} \left(\frac{k_2^2}{k_2^2 + Q_c^2 C^2} - r \right)$$

$$\frac{dP}{dt_1} = \hat{v} \left(1 - \frac{k_4 \alpha}{k_4 + Q_c C} \right) - \hat{\beta} P,$$

As $\epsilon \to 0$, system (2.32) tends to a singular limit, usually called the *reduced system*. We can regard equations (2.30) as a perturbation of the singular limit, resulting from the choice $\epsilon = 0.01 (= \delta)$ in equations (2.32). Alternatively, one can rewrite these equations using a fast time scale, $t = t_1/\epsilon$, which yields

$$\frac{dC}{dt} = \bar{J}_{\text{release}} - \bar{J}_{\text{serca}} + \delta(\bar{J}_{\text{in}} - \bar{J}_{\text{pm}})$$

$$\frac{dC_t}{dt} = \epsilon \bar{J}_{\text{in}} - \bar{J}_{\text{pm}} \tag{2.33}$$

$$\frac{dr}{dt} = \epsilon \frac{1}{\hat{\tau}} \left(\frac{k_2^2}{k_2^2 + Q_c^2 C^2} - r \right)$$

$$\frac{dP}{dt} = \epsilon \hat{v} \left(1 - \frac{k_4 \alpha}{k_4 + Q_c C} \right) - \hat{\beta} P,$$

Equations (2.32) and (2.33) are equivalent for $\epsilon \neq 0$, but taking the limit as $\epsilon \to 0$ of equations (2.33) produce a different singular system, known as the *fast subsystem*.

In the case that a model has two well-separated time scales, GSPT allows one to make predictions about the nature of oscillations occurring in the model, based on knowledge of the dynamics of the reduced system and the fast subsystem. The idea is to construct a singular periodic orbit consisting of alternating fast and slow segments; fast segments are solutions to the fast subsystem and slow segments are solutions of the reduced system. In the simplest cases (including the case where there is just one slow variable) a singular periodic orbit perturbs in a straightforward way when $\epsilon \neq 0$ to produce a relaxation oscillation (RO) in the full system, with the corresponding time series consisting of sections of slow change interspersed with sharp transitions as shown in Fig 6B. In other cases, the singular periodic orbit may perturb to a more complicated orbit such as a mixed mode oscillation (MMO), where the transition from slow to fast segments is via a series of subthreshold oscillations as shown in Fig 6D. The pattern of subthreshold oscillations within an MMO can be quite complicated but can often be predicted using GSPT [34]. A detailed study of the utility of GSPT for the analysis of a variety of different models of intracellular calcium dynamics is contained in [62].

An advantage of the GSPT approach is that the reduced system and the fast subsystem are both effectively of lower dimension than the full system, and so their analysis can be more straightforward than analysis of the full system directly. On the other hand, while GSPT can result in mathematically rigorous results accompanied by appropriate caveats about the regimes of validity of the results, this is not always useful in terms of understanding the dynamics of a model. A common problem is the lack of clear separation between time scales in the model. For instance, in equations (2.30), if $\hat{\tau}$ is $O(10^{-1})$ then the r variable is neither as fast as C nor as slow as C_t and P, and there is not enough of a separation between the speeds of evolution of r and the other variables to define a new intermediate time scale. In such cases, the model might be regarded as being a large perturbation of a singular limit, but then predictions based on a singular limit may be unhelpful. Even if there is clear separation between time scales in a model, there may be more than two time scales present, a situation about which there is little theory.

Some discussion of these kinds of difficulties in the context of calcium models is contained in [62]. One pragmatic approach is to consider a variety of different singular limits. For example to understand the dynamics of equations (2.30) in the case that r is intermediate in speed, one might look at two different limiting cases: one with two fast and two slow variables (with r treated as a fast variable) and the other with one fast and three slow variables (with r treated as a slow variable). One or other of these limiting cases might provide insight into the dynamics of the model, even if neither is close enough to the original model for predictions to be mathematically justified.

One final comment is in order about the use of singular limits in the analysis of calcium models; care is necessary in the identification and analysis of singular limits if misleading results are to be avoided. For instance, the closed cell version of the combined model arises naturally by letting the variable c_t get slower and slower. It is tempting therefore to regard the closed cell model as a singular limit (fast subsystem) of the open cell model, and, by analogy with the procedure followed in GSPT, to assume that the dynamics of the open cell model will be a smooth perturbation of the dynamics of the closed cell model. While some features of the dynamics do perturb in this simple manner, there is a trap: the dynamics of the limit system need not be the same as the limit of the dynamics of the full system. For example, a Hopf bifurcation may be subcritical in the fast subsystem but supercritical in the full system, *no matter how close the full system is to the limiting case*. This issue is discussed in more detail in [157]. A second reason for care in using the closed cell version of the combined model is implicit in the time scale analysis discussed above: in the open cell model, c_t appears to evolve on the same time scale as p (and, possibly, r) in the regime of interest, and so a singular limit in which the speed of evolution of c_t alone (not p or r) tends to zero may not be helpful.

3.2   Pulse experiments and GSPT

An open question for many cell types is whether Ca^{2+} oscillations are principally due to Class I mechanisms (and occur when IP$_3$ concentration is constant), or result from Class II mechanisms (being caused by the intrinsic dynamics of the IPR). One might be tempted to think that, since it is now possible to measure [IP$_3$] and [Ca^{2+}] simultaneously in some cell types [134], this question is easily answered. However, this would not be true. For one thing, these are very difficult experiments to perform, particularly in real cells as opposed to cell lines. Thus, there are still few such measurements in the literature. Secondly, even when one measures [IP$_3$] and [Ca^{2+}] simultaneously not all such questions are immediately answered. For instance, in some cell types, the relative timings of the peak [IP$_3$] and [Ca^{2+}] seem to indicate that a Class I mechanism is required, even though oscillations in [IP$_3$] are observed. In such cases, a peak of [IP$_3$] will naturally follow a peak in [Ca^{2+}] (as Ca^{2+} stimulates the production of IP$_3$) but is not actually necessary for the oscillatory behaviour. For these reasons, it is important to develop additional experimental methods that can be used to distinguish between Class I and Class II mechanisms.

It was proposed in [130] that a simple experiment, involving applying a single exogenous pulse of IP$_3$ to a cell, could be used to determine which type of mechanism was predominant in that cell. The proposal was based on the observation that Class I and Class II models typically respond to a pulse of IP$_3$ in different ways. Specifically, after a pulse of IP$_3$, a Class I model will typically respond with a temporary increase in oscillation frequency while a Class II model will respond with a phase lag, with the next peak in calcium concentration occurring after a delay.

Fig. 9 shows some responses of the rescaled combined model given by equations (2.32) to a pulse of IP$_3$. As in [35] and [61], we model the pulsing process by adding

$$S(t_1) = \hat{M} H(t_1 - t_0) H(t_0 + \Delta - t_1) \tag{2.34}$$

to the right-hand side of the equation for P in the combined model, where $\hat{M}$ denotes the pulse magnitude and H is the Heaviside function

$$H(x) = \begin{cases} 0 \text{ if } x < 0, \\ 1 \text{ if } x \geq 0. \end{cases}$$

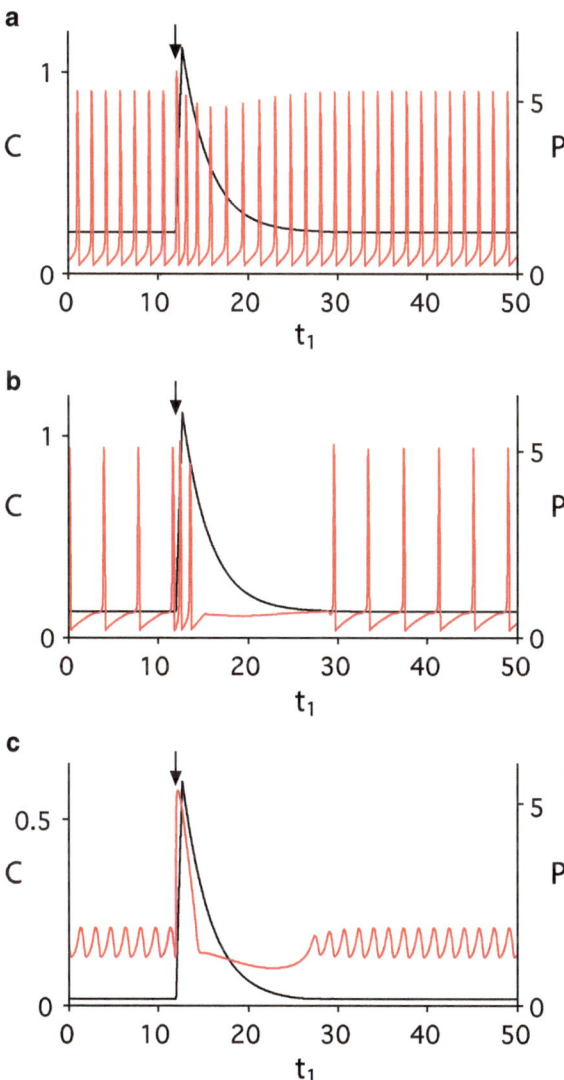

Fig. 9 Responses of equations (2.32) with $\epsilon = 0.01$ to IP$_3$ pulses. The IP$_3$ pulse is applied at the time indicated by the arrow, with the explicit form of the pulse given by equation (2.34) with $\hat{M} = 8.33\dot{3}$, $t_0 = 12$ and $\Delta = 0.72$, assuming that any transients have died out before the time trace is started. Each panel shows the time series of the concentrations C of calcium (red curve) and P of IP$_3$ (black curve). A. Class I: $\alpha = 0$, $\hat{\tau} = 0.48$ ($\tau = 2\,s$) for $\hat{v} = 0.40$ ($v = 0.96$) and other parameter values as in Table 1. B. Class I: as in panel A except with $\hat{v} = 0.233$ ($v = 0.56$). C. Class II: $\alpha = 1$, $\hat{\tau} = 0$ ($\tau = 0$) for $\hat{v} = 0.417$ ($v = 1.00$). Figure modified from [35]

Panel A shows the Class I model response when $\hat{v} = 0.40$ and panel C shows the Class II model response when $\hat{v} = 0.417$. In both cases, the response is the typical case as described above. However, it turns out that there are situations in which a Class I model responds like a Class II model, with a small number of faster oscillations followed by a long quiescent period before oscillations resume. Panel B of Fig. 9 shows this type of response for the Class I version of the combined model when $\hat{v} = 0.233$. The possibility of this anomalous type of response makes interpretation of experimental data ambiguous.

Attempts to understand the anomalous response of some Class I models began in [35], which considered pulse responses for the combined model of section 2.4. The analysis started with the assumption that there was one slow variable in the model, c_t, and used ideas based on the "frozen system" approach, discussed above, to explain the observed dynamics, but the explanation was somewhat ad hoc. The model was re-examined in [61], where it was argued that a comprehensive explanation of the phenomenon required methods from GSPT, and, in particular, that it was necessary to treat the Class I version of the model as a system with three slow variables.

More precisely, [61] worked with the non-dimensionalised Class I combined model given by equations (2.32) with $\alpha = 0$ and $\hat{\tau} = 0.48$, and constructed singular periodic orbits by combining information from the reduced system and the fast subsystem, as described in section 3.1. Specifically, taking the limit $\epsilon \to 0$ of equations (2.32) yields the reduced system, in which the variables evolve on a three-dimensional surface (the *critical manifold*) defined by setting the right-hand side of the dC/dt equation equal to zero. The critical manifold is plotted in Fig. 10 relative to the C, C_t and r coordinates for the case $\hat{v} = 0.317$ and with fixed $P = 0.95$. As can be seen, the critical manifold has two folds relative to the C coordinate. These folds are denoted by blue curves in Fig. 10 and correspond to two-dimensional subsets of the three-dimensional critical manifold in the full four-dimensional phase space. A typical singular orbit of the Class I Atri model then might start on the upper branch of the critical manifold (labelled S_a^+ in Fig. 10), move (slowly) towards the

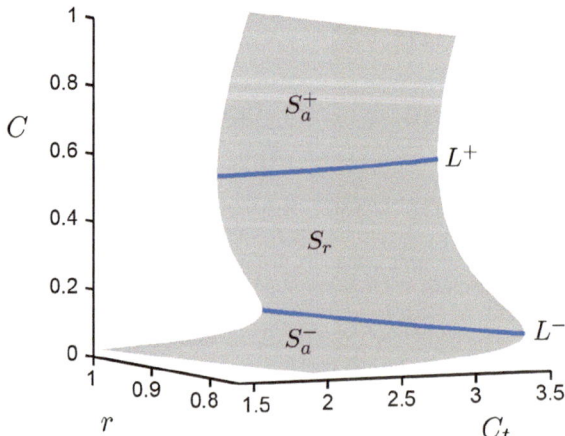

Fig. 10 The critical manifold of the Class I Atri model, equations (2.32) with $\alpha = 0$, $\hat{\tau} = 0.48$, $\hat{v} = 0.317$ and with fixed $P = 0.95$. The surface is divided into three branches (labelled $S_a^{\pm}$ and S_r) by the folds L^- and L^+. Figure modified from [61]

upper fold, make a fast jump to the lower branch of the critical manifold (S_a^-), move slowly towards the lower fold, and then make a second fast jump back to S_a^+. For many parameter values, this singular orbit perturbs when $\epsilon \neq 0$ to an RO or MMO, just as discussed in section 3.1.

Up to this point, the GSPT analysis of the Class I Atri model is fairly standard, but the story becomes more complicated when trying to explain the response of the model to pulsing in IP_3. It was shown in [61] that within the two-dimensional surface of fold points there is a one-dimensional curve of distinguished fold points, called *folded singularities*, that can strongly influence the pulse response of orbits. In certain parameter regimes, pulsed orbits of the full system pass near to the position in phase space at which folded singularities would lie in the singular system; if these folded singularities are of *folded saddle* or *folded node* subtype, a delay in the resumption of oscillations is seen, but if the pulsed orbit stays away from folded saddles or nodes no such delay is observed. Further detail about the analysis of the Class I model is contained in [61], with summary information about folded singularities being given in the review article [34] and the extension of the theory to the case of relevance here (i.e., a system with one fast and three slow variables) being presented in [150].

The next step to understanding the pulse response of the combined model was to look at the Class II model. GSPT methods were used in [61] to show that an unrelated mechanism is responsible for the delay in the pulse response of the Class II model. It was shown that pulsing the Class II model typically sends orbits into a region of phase space where the critical manifold is not folded, meaning that oscillations of the type seen in the Class II model without pulsing (i.e., ROs) are not possible. The pulsed orbit has to spend some time, corresponding to the observed phase lag, travelling back to the region of phase space where the critical manifold is folded before oscillations can resume.

This example is a nice illustration of the power of GSPT in explaining the dynamics of calcium models: the simplest approach, which assumed there is just one slow variable, was not able to properly explain the observations, and a rigorous approach using GSPT was needed. This example also provides an instance in which physiological considerations (i.e., the desire to explain the pulse responses) stimulated the development of new mathematics (e.g., the extension of GSPT to the case of three or more slow variables [150]).

4 Merging calcium dynamics and membrane electrical excitability

Many of the techniques used in the study of Ca^{2+} oscillations were developed in studies of the generation of oscillatory action potentials in neurons and other excitable cells. The membrane potential is by far the best known, and most widely studied, cellular oscillator, with most theoretical work based ultimately on the 1952 model of Hodgkin and Huxley [65]. It is far beyond the scope of the present work to discuss membrane potential models in detail; introductions to the theoretical study of membrane oscillators can be found in [73, 76].

However, no discussion of Ca^{2+} oscillations would be complete without at least a brief mention of how they interact with membrane potential oscillators. As a general rule, oscillations in the membrane potential (usually taking the form of oscillatory spiking) are caused by oscillatory opening and closing of ion channels (typically Na^+, K^+ or Ca^{2+} channels) in the cell membrane. Such oscillations in the membrane potential typically occur on a millisecond time scale, orders of magnitude faster than the Ca^{2+} oscillations discussed here.

However, many cells have ion channels whose conductances are controlled by $[Ca^{2+}]$. In this case, slow oscillations in $[Ca^{2+}]$ can be used to modulate, over a longer time scale, the properties of the fast electrical oscillation. For example, slow oscillations in $[Ca^{2+}]$ can move the membrane potential model in and out of the oscillatory regime (by, say, slow modulation of the K^+ conductance), resulting in bursts of action potentials, a phenomenon known as electrical bursting, and seen in a wide variety of neurons and neuroendocrine cells. The paper by Bertram et al. in this volume presents a detailed discussion of one such type of model. Other examples can be found in [64, 68], while a basic introduction to the field can be found in [73].

Such systems, which couple a slower cytosolic Ca^{2+} oscillator to a faster membrane potential oscillator, have the potential for a wide range of complex and interesting dynamical behaviours. From a mathematical point of view, the complexity may arise, in part at least, from the multitude of time scales involved; models of membrane potential oscillators typically have at least two time scales, and calcium oscillator models also typically have at least two time scales, so combined models will typically have three or more time scales, depending on the relative speeds of the slower variable(s) in the membrane potential model and the faster variable(s) in the calcium model. A comprehensive theory of dynamics in systems with more than two time scales has yet to be developed, but early work indicates that very complex phenomena can occur in this context [79, 80]. From a physiological view point, models that couple a cytosolic Ca^{2+} oscillator to a faster membrane potential oscillator have particular importance in the study of neuroendocrine cells [9, 10, 11, 49, 82, 84, 87, 110, 144, 143, 156], and thus in the study of hormonal control, and are sure to be a major area of mathematical and experimental research in the future.

5   Calcium diffusion and waves

5.1   Basic equations

To turn a simple spatially homogeneous model into a model that allows for a spatially varying $[Ca^{2+}]$ (as is, of course, the case in reality), the model equations must be adapted to include the diffusion of Ca^{2+}, and this requires, in practice, a host of additional assumptions.

Firstly, rather than modelling the ER and the cytoplasm as two distinct spaces, connected by Ca^{2+} fluxes, it is sufficient for most applications to combine these regions into a single homogenised domain, in which the ER and the cytoplasm co-exist at every point in space, and Ca^{2+} within each space has an effective diffusion coefficient that depends on the exact geometry assumed in the homogenisation [54]. Thus, we get the following equations for evolution of c and c_e:

$$\frac{\partial c}{\partial t} = \nabla \cdot (D_c^{\text{eff}} \nabla c) + \chi_c f(c, c_e), \tag{2.35}$$

$$\frac{\partial c_e}{\partial t} = \nabla \cdot (D_e^{\text{eff}} \nabla c_e) + \chi_e g(c, c_e), \tag{2.36}$$

where D_c^{eff} and D_e^{eff} are effective diffusion coefficients for the cytoplasmic space and the ER, respectively, χ_c and χ_e are the surface-to-volume ratios of these two co-mingled spaces, and $f(c, c_e)$ and $g(c, c_e)$ denote all the other Ca^{2+} fluxes and reactions.

It is usually assumed that the cellular cytoplasm is isotropic and homogeneous. It is not known, however, how Ca^{2+} diffuses in the ER, or the extent to which the tortuosity of the ER plays a role in determining the effective diffusion coefficient of ER Ca^{2+}. Thus, it is typical (and reasonable) to assume either that Ca^{2+} does not diffuse in the ER, or that it does so with a restricted diffusion coefficient, $D_e^{\text{eff}} \ll D_c^{\text{eff}}$. Henceforth we delete the superscript eff.

In this case, the simplified equations for Ca^{2+} diffusion are

$$\frac{\partial c}{\partial t} = D_c \nabla^2 c + f(c, c_e) + k_- b - k_+ c(b_t - b), \tag{2.37}$$

$$\frac{\partial c_e}{\partial t} = D_e \nabla^2 c_e + g(c, c_e), \tag{2.38}$$

$$\frac{\partial b}{\partial t} = D_b \nabla^2 b - k_- b + k_+ c(b_t - b), \tag{2.39}$$

where χ_c and χ_e have been absorbed into the other model parameters, and where cytoplasmic Ca^{2+} buffering has been explicitly included, for reasons that will become clear soon. ER Ca^{2+} buffering is not included explicitly, purely for simplicity. To do so makes no difference to the analysis, it merely makes the notation more complex.

As in the absence of diffusion, when buffering is fast the model can be condensed [122, 123, 147]. Assuming, as before, that

$$k_- b - k_+ c(b_t - b) = 0, \tag{2.40}$$

we get the "slow" equation

$$\frac{\partial}{\partial t}(c + b) = D_c \nabla^2 c + D_b \nabla^2 b + f(c, c_e), \tag{2.41}$$

which, after eliminating b, becomes

$$\frac{\partial c}{\partial t} = \frac{1}{1 + \theta(c)} \left(\nabla^2 \left(D_c c + D_b b_t \frac{c}{K + c} \right) + f(c, c_e) \right) \tag{2.42}$$

$$= \frac{D_c + D_b \theta(c)}{1 + \theta(c)} \nabla^2 c - \frac{2 D_b \theta(c)}{(K + c)(1 + \theta(c))} |\nabla c|^2 + \frac{f(c, c_e)}{1 + \theta(c)}, \tag{2.43}$$

where, as before,

$$\theta(c) = \frac{b_t K}{(K + c)^2}. \tag{2.44}$$

Note that we assume that b_t does not vary in either space or time. A similar equation holds for c_e.

Nonlinear buffering changes the model structure significantly, although it can have surprisingly little qualitative effect on the resulting dynamics [52, 140]. In particular, Ca^{2+} obeys a nonlinear diffusion–advection equation, where the advection is the result of Ca^{2+} transport by a mobile buffer. The effective diffusion coefficient

$$D_{\text{eff}} = \frac{D_c + D_b \theta(c)}{1 + \theta(c)} \tag{2.45}$$

is a convex linear combination of the two diffusion coefficients D_c and D_b, so lies somewhere between the two. Since buffers are large molecules, $D_{\text{eff}} < D_c$. If the buffer is not mobile, i.e., $D_b = 0$, then (2.43) reverts to a reaction–diffusion equation. Also, when Ca^{2+} gradients are small, the nonlinear advective term can be ignored.

If the buffer is not only fast, but also of low affinity, so that $K \gg c$, then θ is constant, and D_{eff} is constant also.

It is commonly assumed that the buffer has fast kinetics, is immobile, and has a low affinity. With these assumptions we get the simplest possible model of Ca^{2+} buffers (short of not including them at all), in which

$$\frac{\partial c}{\partial t} = \frac{K}{K + b_t} (D_c \nabla^2 c + f(c)), \tag{2.46}$$

wherein both the diffusion coefficient and the fluxes are scaled by the constant factor $K/(K + b_t)$; each flux in the model can then be interpreted as an *effective* flux, i.e., that fraction of the flux that contributes to a change in free Ca^{2+} concentration.

5.2 Fire-diffuse-fire models

One particularly simple way in which calcium excitability can be used to model waves is with the fire-diffuse-fire model [32, 75, 25, 28, 29], a direct analogue of the spike-diffuse-spike model of action potential propagation [26, 27]. In this model, once [Ca^{2+}] reaches a threshold value, c^*, at a release site, that site fires, instantaneously releasing a fixed amount, σ, of Ca^{2+}. Thus, a Ca^{2+} wave is propagated by the sequential firing of release sites, each responding to the Ca^{2+} diffusing from neighbouring release sites. Hence the name fire–diffuse–fire.

In the fire-diffuse-fire model Ca^{2+} obeys the reaction–diffusion equation

$$\frac{\partial c}{\partial t} = D_c \frac{\partial^2 c}{\partial x^2} + \sigma \sum_n \delta(x - nL)\delta(t - t_n), \tag{2.47}$$

where L is the spacing between release sites. Although this equation looks linear, appearances are deceptive. Here, t_n is the time at which c first reaches the threshold value c^* at the nth release site, and thus depends in a complicated way on c.

The Ca^{2+} profile resulting from the firing of a single site, site i, say, is

$$c_i(x, t) = \sigma \frac{H(t - t_i)}{\sqrt{4\pi D_c(t - t_i)}} \exp\left(-\frac{(x - iL)^2}{4D_c(t - t_i)}\right), \tag{2.48}$$

where H is the Heaviside function. This is the fundamental solution of the diffusion equation with a delta function input at $x = i, t = t_i$. If we superimpose the solutions from each site, we get

$$c(x, t) = \sum_i c_i(x, t) = \sigma \sum_i \frac{H(t - t_i)}{\sqrt{4\pi D_c(t - t_i)}} \exp\left(-\frac{(x - iL)^2}{4D_c(t - t_i)}\right). \tag{2.49}$$

Notice that because of the instantaneous release, $c(x, t)$ is not a continuous function of time at any release site.

From this explicit expression it is possible to calculate an explicit expression for the wave speed. For full details the reader is referred to the abbreviated discussion in [73] or the more detailed presentations in the original articles referenced above.

This version of the fire-diffuse-fire model has no Ca^{2+} removal, and thus the concentration of Ca^{2+} is always increasing. This can be remedied by the inclusion of a Ca^{2+} removal term [25], modelling the removal by SERCA pumps. However, in order to preserve the analytical tractability of this approach, the removal term must be linear.

5.3   Another simple example

To illustrate some of the main features of wave propagation in Ca^{2+} models, we use a model similar to the combined model of Section 2.4, but somewhat simpler. Firstly, we include diffusion in one spatial dimension only. Even though Ca^{2+} waves propagate in three dimensions, a model in one spatial dimension is not necessarily a bad approximation. Since the wavelength of a typical Ca^{2+} wave is large compared to the dimensions of a typical cell, much (but not all) intracellular wave propagation is essentially one-dimensional in nature. It is only when one considers wave propagation in much larger cells, such as a Xenopus oocyte, that the two and three dimensional properties of the waves become apparent, as the waves form spirals and target patterns [83].

We make a number of additional simplifications. Firstly, we assume that the Ca^{2+} ATPase pumps are linearly dependent on $[Ca^{2+}]$. Since we know this to be untrue, our simplified model will never be a good quantitative description of real Ca^{2+} waves. However, much of the underlying dynamical behaviour is preserved by this assumption. Secondly, we assume that the flux through the IPR is a bell-shaped function of $[Ca^{2+}]$, with no time delays. Hence, our simplified model here is neither a Class I nor a Class II model. In this case, the oscillations in $[Ca^{2+}]$ are entirely dependent on Ca^{2+} influx from the outside. Although this is the case in only some cell types, the model still serves to illustrate the basic dynamical properties of Ca^{2+} models.

With these assumptions, our model equations are

$$\frac{\partial c}{\partial t} = D_c \frac{\partial^2 c}{\partial x^2} + J_{IPR} - k_s c + \varepsilon(J_{influx} - k_p c),$$

$$\frac{\partial c_e}{\partial t} = \gamma(-J_{IPR} + k_s c), \tag{2.50}$$

where

$$J_{influx} = k_{in} p, \tag{2.51}$$

$$J_{IPR} = \left(\alpha + k_f p \left(\frac{c^2}{c^2 + \varphi_1^2}\right)\left(\frac{\varphi_2}{\varphi_2 + c}\right)\right)(c_e - c). \tag{2.52}$$

As before, p denotes $[IP_3]$, and is treated as the principal bifurcation parameter. The expression for J_{influx} is merely a slightly simplified version of equation (2.27). Because oscillations in this model depend on Ca^{2+} entry and exit from the cell, it is also possible to let J_{influx} be a parameter, and use it as the principal bifurcation parameter [142]. Typical values of the other model parameters are given in Table 2 in the Appendix.

5.4 CU systems

A convenient first step in investigating wave propagation in PDE models of calcium dynamics is to switch to a moving frame. For a model with one spatial variable, x, with solitary or periodic waves moving with a constant wave speed s, we can define a new variable, $z = x + st$, and rewrite the model in the moving frame. For instance, in terms of this new variable, the model given by equations (2.50) becomes:

$$c' = u,$$

$$u' = \frac{1}{D_c}\left(su - J_{\mathrm{IPR}} + k_s c - \varepsilon(J_{\mathrm{influx}} - k_p c)\right),$$

$$c_e' = \frac{\gamma}{s}(-J_{\mathrm{IPR}} + k_s c), \tag{2.53}$$

where the prime denotes differentiation with respect to z.

We are interested in both pulse-type travelling waves and periodic travelling waves for the PDE model; in the moving frame ODEs, these correspond, respectively, to homoclinic orbits and periodic solutions. Typically, we will be interested in the existence of such solutions as both a bifurcation parameter of the PDE (e.g., p for the model above) and s, the wave speed, vary. In the PDE formulation, s is a quantity selected by the dynamics, not a parameter of the equations, but in the travelling wave ODEs we treat s as a bifurcation parameter.

A first step is to look for homoclinic and Hopf bifurcations of the moving frame ODEs in the corresponding two-dimensional parameter space. For example, for the parameter values specified in Table 2, equations (2.53) have a unique equilibrium point, which is of saddle type with a one-dimensional unstable manifold and a two-dimensional stable manifold for p and s values outside the U-shaped curve labeled HB in Fig. 11. This equilibrium has a homoclinic bifurcation at (p, s) values on the C-shaped curve (labelled HC) in this figure.

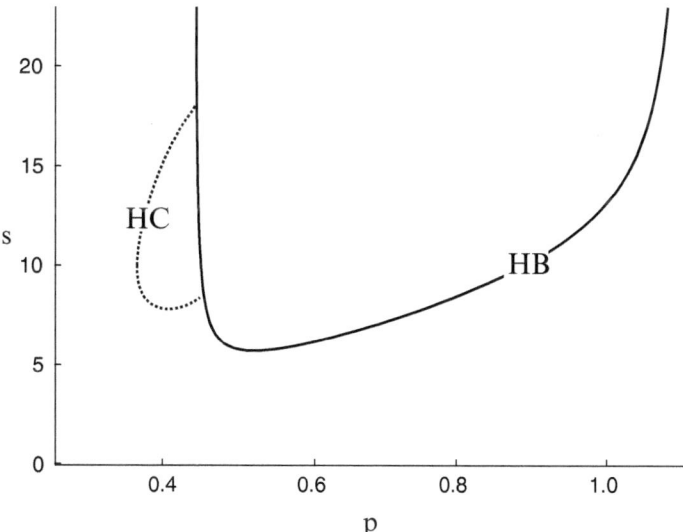

Fig. 11 Partial bifurcation set for equations (2.53) for parameter values given in Table 2, showing a U-shaped curve of Hopf bifurcations (HB) and a C-shaped dotted curve of homoclinic bifurcations (HC)

   The structure observed in Fig. 11, of a C-shaped homoclinic bifurcation curve and a U-shaped Hopf locus, turns out to be common to many models of calcium waves, as well as many other excitable systems such as the FitzHugh-Nagumo and Hodgkin-Huxley models [20]. It is argued in [20] and [142] that the CU-structure occurs as a consequence of the general shape of the nullclines in these models, which in turn follows from the underlying physiology. Furthermore, Maginu [90] showed that in the limit of $s \to \infty$, the travelling wave equations reduce to the model without diffusion (i.e., with $D_c = 0$); since the diffusion-free version of a calcium model will typically have two Hopf bifurcations at finite values of the bifurcation parameter (as discussed in section 3), this result suggests that the Hopf locus really is U-shaped, i.e., the left and right arms of the Hopf locus will have vertical asymptotes at finite values of the bifurcation parameter.

   For each fixed value of the main bifurcation parameter between the vertical asymptotes of the Hopf U there will typically be an interval of s values for which periodic solutions exist. It is natural to ask which of these periodic solutions will give stable periodic travelling waves in the PDE, i.e., to ask which wave speed will be selected by the PDE dynamics. There is no known general answer to this question; the answer is believed to depend on the precise boundary and initial conditions for the PDE. It is known [119] that very complicated, seemingly chaotic, travelling solutions can occur at values of the bifurcation parameter lying within the Hopf U.

   Analysis of the moving frame ODEs can tell us about the existence of travelling waves in the associated PDE model, but does not give information about stability of these solutions in the PDEs. Instead, stability of travelling waves can be determined by direct computation (e.g., [109]) or by numerical computation on the PDEs (e.g., [119, 142]). In all the cases we have studied, it turns out that stable solitary travelling waves have wave speeds corresponding to the upper 'branch' (higher s values) of the C curve, although this branch may not be stable along its entire length. More complicated travelling pulses (e.g., with two pulses within the wave packet, corresponding to double-pulse homoclinic orbits in the travelling wave equations) may also occur [20] and can be stable [109].

   Just as for ODE models, PDE models of calcium dynamics typically have processes occurring on two or more different time scales, and it is possible to exploit this time scale separation to explain model dynamics. Such ideas have been very successful in the analysis of the PDE version of the FitzHugh-Nagumo equations (e.g., [3, 72, 78]), but have been applied less to calcium models. One approach has been to look for the singular analogue of the CU structure, and to try to show that features of the bifurcation set of the full (non-singular) problem, including the CU structure, arise as perturbations of this singular structure. In [142], the existence and stability of travelling waves in a closed-cell (singular) version of a calcium model closely related to equations (2.53) was investigated theoretically, and the results compared with numerical results for the (non-singular) open-cell model. It was shown that the CU structure for the full system, found numerically, appears to converge in the singular limit to a collection of fronts, pulses and waves that can be located analytically in the singular limit system and that form a singular CU structure. Work is underway to show rigorously how the singular CU structure perturbs to the nonsingular case.

Although the basic CU structure is common to many calcium models, other features of the bifurcation set vary from model to model. For instance, different models may exhibit a variety of different types of global bifurcations, including homoclinic and heteroclinic bifurcations of equilibria and periodic orbits [158], and give rise to a host of interesting issues from a bifurcation theory point of view, but these are not our focus in this article. One aspect of the dynamics of particular interest is how the C curve terminates near its apparent endpoints; this question has implications for the ways in which there can be a transition from stable travelling pulses to stable periodic travelling waves in the PDE and was discussed in [20, 119] for some specific models.

5.5   Calcium excitability and comparison to the FitzHugh-Nagumo equations

A crucially important feature of models of Ca^{2+} waves is excitability; a small amount of Ca^{2+} release induces the release of a larger amount of Ca^{2+} through positive feedback in the model. The most studied excitable system is the FitzHugh-Nagumo equations, and it has long been recognised that calcium waves propagate by an excitable mechanism similar in many ways to that in the FitzHugh-Nagumo model. Despite these similarities, however, there are important differences.

The FitzHugh-Nagumo equations can be written in the form

$$\frac{\partial u}{\partial t} = D\frac{\partial^2 u}{\partial x^2} + u(u - \alpha)(1 - u) - w + I,$$

$$\frac{\partial w}{\partial t} = \epsilon(u - \gamma w), \tag{2.54}$$

where the variable u represents the plasma membrane electric potential, w represents the combined inactivation effects of the sodium and potassium channels, and I is the applied current. The parameter ϵ satisfies $0 \leq \epsilon \ll 1$, and encodes the separation of time scales in the model, $\alpha \in (0, \frac{1}{2})$, D is the diffusion constant and γ is a small positive constant [73]. Defining $z = x + st$ in the usual way, where s is the wave speed, yields the model equations in the moving frame:

$$\frac{du}{dz} = v,$$

$$\frac{dv}{dz} = \frac{1}{D}\left(sv - u(u - \alpha)(1 - u) + w - I\right), \tag{2.55}$$

$$\frac{dw}{dz} = \frac{\epsilon}{s}(u - \gamma w).$$

In the absence of diffusion the dynamics of the FitzHugh-Nagumo equations and the dynamics of typical calcium models, such as equations (2.50), are qualitatively very similar. Any difference in dimension of the models (if the calcium model has three or more dependent variables) gives different possibilities for the detailed dynamics, but structural similarities in the models, specifically a clearly defined slow variable such as w for FitzHugh-Nagumo and c_t for calcium models, and the cubic shape of the nullcline for a fast variable (v for FitzHugh-Nagumo and c for calcium models), results in the diffusion-free models having similar bifurcation diagrams and time series. For instance, Fig. 12 shows bifurcation diagrams and typical time series for equations (2.50) with $D_c = 0$ and for equations (2.54) with $D = 0$; the similarities in the model dynamics are clear in these pictures.

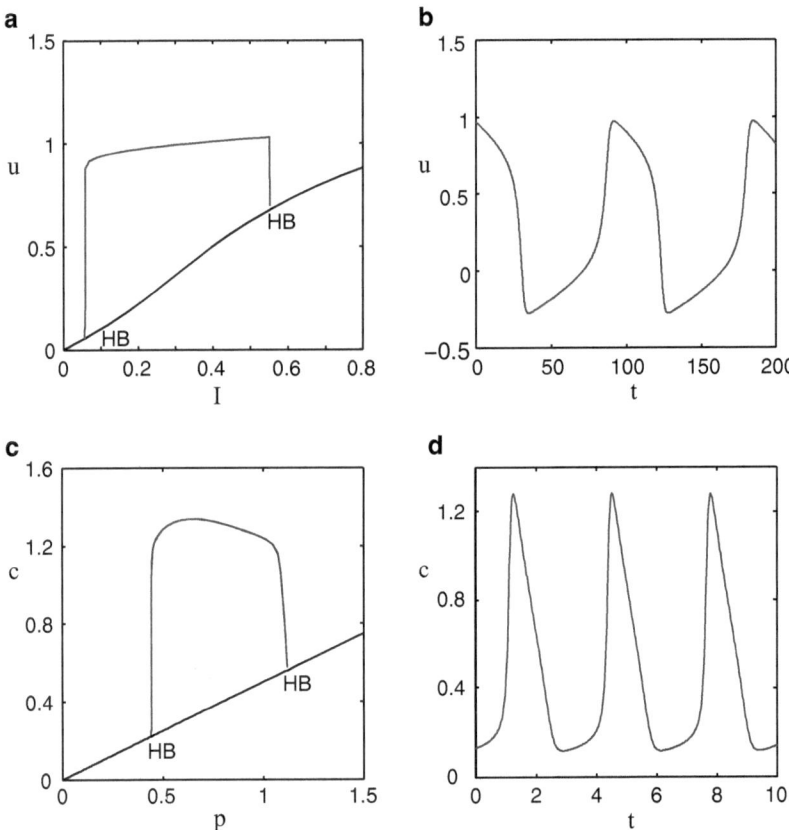

Fig. 12 Bifurcation diagrams and time series for the FitzHugh-Nagumo model, equations (2.54), and a simple calcium model, equations (2.50) without diffusion. Panel A shows the bifurcation diagram for equations (2.54) with $D = 0$, $a = 0.1$, $\gamma = 1.0$, $\epsilon = 0.1$. The black curve shows the position of the steady state solution and the blue curve indicates the maximum amplitudes of periodic orbits. Hopf bifurcations are labelled HB. Panel B shows the time series for the corresponding attracting periodic solution when $I = 0.2$. Panel C shows the bifurcation diagram for equations (2.50) with $D_c = 0$ and other parameter values as in Table 1. Line styles and labels as for panel A. Panel D shows the time series for the corresponding attracting periodic solution when $p = 0.7$

If diffusion is included, then there are still marked similarities between the dynamics of the FitzHugh-Nagumo equations and a typical calcium model. Most notably, the FitzHugh-Nagumo moving frame equations have a CU bifurcation structure in the (I, s) parameter plane very similar to that in calcium models [20, 142]. However, structural differences in the models mean that the underlying mechanisms can be quite different.

One important structural difference results from the way diffusion acts. In the FitzHugh-Nagumo model, diffusion appears in the evolution equation for the fast variable, u, only. This follows from a modelling assumption that the gating variable, w, is uniformly distributed along the spatial direction. The situation is, typically, different for calcium models, where diffusion affects both fast and slow variables since diffusion in the cytoplasm influences the evolution of both c, the cytoplasmic calcium concentration, and c_t, the total cellular concentration.

To see this in the case of equations (2.50) we go to the moving frame by setting $z = x + st$ and $u = dc/dz$, then rewrite the model in standard fast-slow form by replacing c_e with $c_t = s(c_e/\gamma + c) - D_c u$. This definition of c_t is the PDE analogue of the total calcium variable introduced in section 2.4. With these changes, equations (2.53) become:

$$c' = u,$$

$$u' = \frac{1}{D_c}\left(su - \bar{J}_{\text{IPR}}\left(\frac{\gamma}{s}(c_t + D_c u - sc) - c\right) + k_s c - \varepsilon(J_{\text{influx}} - k_p c)\right),$$

$$c_t' = \varepsilon(J_{\text{influx}} - k_p c), \tag{2.56}$$

where the prime indicates differentiation with respect to z and

$$\bar{J}_{\text{IPR}} = \alpha + k_f p\left(\frac{c^2}{c^2 + \varphi_1^2}\right)\left(\frac{\varphi_2}{\varphi_2 + c}\right).$$

When ε is sufficiently small, c and u are fast variables and c_t is slow. One effect of the diffusion of calcium is, therefore, to introduce nonlinear coupling between the fast variables, i.e., a term of the form of $ug(c)$ in the u' equation, for $g(c)$ a nonlinear function of c. By comparison, in equations (2.55) there are no comparable terms in the differential equations for the fast variables.

A direct consequence of this difference is seen in the nature of the Hopf bifurcations. In the FitzHugh-Nagumo equations, the simple coupling between the fast variables means the Hopf bifurcations that occur on the vertical arms of the Hopf U are degenerate in the singular limit, in the sense that the bifurcation is neither supercritical or subcritical because the first Lyapunov coefficient is zero. For more generic coupling, as found in calcium models such as equations (2.50), this is not the case [142], and the Hopf bifurcations on the vertical arms will be either super- or sub-critical as the singular limit is approached. Note that in both types of model, the Hopf bifurcations are *singular Hopf* bifurcations in the singular limit (so that the imaginary parts of the eigenvalues at the Hopf bifurcations tend to zero as $\epsilon \to 0$; see [13]), but this singularity is distinct from the degeneracy arising from simple coupling in the FitzHugh-Nagumo equations. More work is needed to uncover exactly how this difference between the singular versions of the models influences the dynamics of the non-singular models.

A second important structural difference between the models occurs because the FitzHugh-Nagumo equations have a symmetry: equations (2.55) are equivariant with respect to the transformation

$$u \to \frac{2}{3}(1 + \alpha) - u, \quad v \to -v, \quad w \to \frac{2}{3\gamma}(1 + \alpha) - w,$$

$$I \to \frac{2}{3}(1 + \alpha)\left[\frac{1}{\gamma} - \frac{(2 - \alpha)(1 - 2\alpha)}{9}\right] - I.$$

As a consequence, some of the travelling pulses in the FitzHugh-Nagumo model arise as a perturbation of two symmetry-related singular travelling front solutions, corresponding in the moving frame to two symmetry-related heteroclinic orbits [142]. (Other travelling pulses arise as a perturbation of a singular travelling pulse, corresponding to a homoclinic orbit in the moving frame.) By contrast, calcium models typically do not have such a symmetry, and travelling wave solutions are unlikely to arise in this way. In a model closely related to equations (2.50) it appears that the corresponding travelling pulses also arise from the singular limit as a perturbation of travelling fronts [142], but the mechanism is more generic than in the FitzHugh-Nagumo model since it does not require the presence of symmetry. More work is necessary to establish whether this is the usual pattern in calcium models.

In summary, there are important structural differences between the FitzHugh-Nagumo equations and typical calcium models, which arise because of simplifying, non-generic assumptions made in constructing the FitzHugh-Nagumo equations. We conclude that models of calcium dynamics are excitable systems of a somewhat different type to the FitzHugh-Nagumo equations.

5.6   The effects on wave propagation of calcium buffers

Another way in which Ca^{2+} models differ from more widely studied models such as the FitzHugh-Nagumo equations is the presence of buffers. As discussed in Section 5.1, Ca^{2+} buffers effectively disappear from the model equations only under the rather restrictive assumptions of fast, linear, buffering. Since such assumptions are unlikely to be accurate in most cells, it is important to understand the dynamics of wave propagation in the presence of more general nonlinear or slow buffers. If the buffer is fast, but nonlinear, then we still have a single transport equation (equation (2.43)), but if the buffers are slow, then we are forced to deal with an additional equation (equation (2.39)).

The possible effects of buffers on waves are particularly interesting in the context of Ca^{2+} waves, as, experimentally, Ca^{2+} waves are observed by adding Ca^{2+} fluorescent dyes to cells. However, since these dyes are necessarily also Ca^{2+} buffers (as they must bind Ca^{2+} in order to emit light) questions have been raised about how much of the observed behaviour is an artefact of the experimental method. For example, is it possible for waves to exist only in the presence of the additional buffer represented by the dye, or do they exist even when they are not being measured?

There have been a number of studies, both numerical and analytic [69, 81, 99, 111, 122, 123, 121, 147], of the effects of Ca^{2+} buffers. By far the most analytical work on this question has been done by Je-Chiang Tsai [139, 140, 141, 142]. Almost all of this analytic work has been done on the FitzHugh-Nagumo model, the prototypical excitable system, or on the bistable equation, which is merely the FitzHugh-Nagumo model with no recovery variable. In the notation of this paper, the buffered bistable equation is just equation (2.37), with $f(c, c_e) = c(1-c)(c-\alpha)$, for some constant $0 < \alpha < 1/2$; see [73] for an introductory discussion of wave propagation in the bistable and FitzHugh-Nagumo equations. As yet, it is not entirely clear how results from the buffered versions of the bistable or FitzHugh-Nagumo equations carry over to models of Ca^{2+} waves, but since these are the only excitable systems for which any significant amount of analytical work has been done, it is the best we can currently do. In addition, numerical solutions indicate that these analytical results carry over, in most part, to Ca^{2+} waves. Although this is not a proof, of course, it offers some reassurance.

Tsai has shown that, when the buffers are fast, there is a unique, stable, travelling wave solution to the buffered bistable equation, a result entirely analogous to the result for the unbuffered bistable equation. If the buffers are slow and immobile then, again, the same results holds; i.e., there exists a unique, stable, travelling wave solution. It is important to note that these travelling waves, although their existence, uniqueness and stability is guaranteed, may well have quite different forms or profiles from waves in the unbuffered bistable equation.

When the buffers are slow and mobile, the situation is more complicated. It is possible to eliminate waves by the addition of enough slow, mobile, buffer, but, when the waves exist, they are still unique and stable.

The buffered FitzHugh-Nagumo equation is considerably more complicated, having, as it does, an additional equation for the recovery variable, and presently there are analytical results only for the case of fast buffering. In this case there is a complex relationship between the binding constant of the buffer (i.e., the ratio k_-/k_+, which determines how strong the buffer is), the excitability of the system (α), and the time scale separation (ε). If there is too much buffer present, waves will not exist. However, in some conditions, waves can be made to exist by the addition of a small amount (but not too much) of weakly binding buffer. Our current knowledge of the buffered FitzHugh-Nagumo equation is summarised in [141], although many gaps remain.

6 Conclusion

This review has focused on only a very restricted range of models of Ca^{2+} dynamics, but even this small range suffices to show how these models raise a host of important mathematical questions. Not only are these questions proving to be vital for the interpretation of some experimental data, they also have applicability well outside the immediate area of Ca^{2+} dynamics, particularly in the study of membrane potential models, or models of chemical reactions with multiple time scales.

Acknowledgements This work was supported by the Marsden Fund of the Royal Society of New Zealand and by the National Institutes of Health of the USA (NIDCR R01DE019245).

Appendix

Table 1 Values of parameters for the combined model, equations (2.20)–(2.23)

Parameter	Value	Parameter	Value	Parameter	Value
b	0.111	k_2	0.7 μM	k_{flux}	6 s^{-1}
δ	0.01	k_4	1.1 μM	V_p	24.0 μM s^{-1}
γ	5.405	k_p	0.4 μM	V_e	20.0 μM s^{-1}
μ_0	0.567	k_e	0.06 μM	α_1	1.0 μM s^{-1}
μ_1	0.433	k_1	1.1 μM	α_2	0.2 s^{-1}
V_1	0.889	k_μ	4.0 μM	β	0.08 s^{-1}

Table 2 Values of parameters for the model defined by equations (2.50)

α	k_s	k_f	k_p	φ_1	φ_2	ε	γ	k_{in}	D_c
0.05 s^{-1}	20 s^{-1}	20 s^{-1}	20 s^{-1}	2 μM	1 μM	0.1	5	10 s^{-1}	20 μm^{2}s^{-1}

References

[1] Ambudkar IS (2012) Polarization of calcium signaling and fluid secretion in salivary gland cells. Curr Med Chem 19(34):5774–81

[2] Atri A, Amundson J, Clapham D, Sneyd J (1993) A single-pool model for intracellular calcium oscillations and waves in the Xenopus laevis oocyte. Biophys J 65(4):1727–39, DOI 10.1016/S0006-3495(93)81191-3

[3] Bell D, Deng B (2002) Singular perturbation of N-front travelling waves in the FitzHugh–Nagumo equations. Nonlinear analysis: Real world applications 3(4):515–541

[4] Berridge MJ (2009) Inositol trisphosphate and calcium signalling mechanisms. Biochim Biophys Acta 1793(6):933–40

[5] Berridge MJ (2012) Calcium signalling remodelling and disease. Biochem Soc Trans 40(2):297–309

[6] Berridge MJ, Lipp P, Bootman MD (2000) The versatility and universality of calcium signalling. Nat Rev Mol Cell Biol 1(1):11–21

[7] Bers D (2000) Calcium fluxes involved in control of cardiac myocyte contraction. Circ Res 87(4):275–281

[8] Bers DM (2001) Excitation-contraction coupling and cardiac contractile force. Second edition. Kluwer, New York

[9] Bertram R (2005) A mathematical model for the mating-induced prolactin rhythm of female rats. AJP: Endocrinology and Metabolism 290(3):E573–E582

[10] Bertram R, Previte J, Sherman A, Kinard TA, Satin LS (2000) The phantom burster model for pancreatic b -cells. Biophys J 79(6):2880–2892

[11] Bertram R, Sherman A, Satin LS (2010) Electrical bursting, calcium oscillations, and synchronization of pancreatic islets. Adv Exp Med Biol 654:261–79

[12] Bootman M, Niggli E, Berridge M, Lipp P (1997) Imaging the hierarchical Ca^{2+} signalling system in HeLa cells. J Physiol 499(2):307–14

[13] Braaksma B (1998) Singular Hopf bifurcation in systems with fast and slow variables. Journal of Nonlinear Science 8:457–490

[14] Brini M, Carafoli E (2009) Calcium pumps in health and disease. Physiol Rev 89(4):1341–1378

[15] Callamaras N, Marchant JS, Sun XP, Parker I (1998) Activation and co-ordination of InsP$_3$-mediated elementary Ca^{2+} events during global Ca^{2+} signals in Xenopus oocytes. J Physiol 509(1):81–91

[16] Cannell MB, Kong CHT (2012) Local control in cardiac E–C coupling. Journal of Molecular and Cellular Cardiology 52(2):298–303

[17] Cannell MB, Soeller C (1999) Mechanisms underlying calcium sparks in cardiac muscle. J Gen Physiol 113(3):373–6

[18] Cao P, Donovan G, Falcke M, Sneyd J (2013) A stochastic model of calcium puffs based on single-channel data. Biophys J 105(5):1133–42, DOI 10.1016/j.bpj.2013.07.034

[19] Catterall WA (2011) Voltage-gated calcium channels. Cold Spring Harb Perspect Biol 3(8):a003,947, DOI 10.1101/cshperspect.a003947

[20] Champneys A, Kirk V, Knobloch E, Oldeman B, Sneyd J (2007) When Shil'nikov meets Hopf in excitable systems. SIAM J Appl Dyn Syst 6:663–693

[21] Cheer A, Nuccitelli R, Oster G, Vincent JP (1987) Cortical waves in vertebrate eggs I: the activation waves. J Theor Biol 124:377–404

[22] Cheng H, Lederer WJ, Cannell MB (1993) Calcium sparks: elementary events underlying excitation-contraction coupling in heart muscle. Science 262(5134):740–4

[23] Colegrove S, Albrecht M, Friel D (2000a) Quantitative analysis of mitochondrial Ca^{2+} uptake and release pathways in sympathetic neurons. reconstruction of the recovery after depolarization-evoked $[Ca^{2+}]_i$ elevations. J Gen Physiol 115:371–88

[24] Colegrove SL, Albrecht MA, Friel D (2000b) Dissection of mitochondrial Ca^{2+} uptake and release fluxes in situ after depolarization-evoked $[Ca^{2+}]_i$ elevations in sympathetic neurons. J Gen Physiol 115(3):351–370

[25] Coombes S (2001a) The effect of ion pumps on the speed of travelling waves in the fire-diffuse-fire model of Ca^{2+} release. Bull Math Biol 63(1):1–20

[26] Coombes S (2001b) From periodic travelling waves to travelling fronts in the spike-diffuse-spike model of dendritic waves. Math Biosci 170(2):155–72

[27] Coombes S, Bressloff PC (2003) Saltatory waves in the spike-diffuse-spike model of active dendritic spines. Phys Rev Lett 91(2):028,102

[28] Coombes S, Timofeeva Y (2003) Sparks and waves in a stochastic fire-diffuse-fire model of Ca^{2+} release. Phys Rev E Stat Nonlin Soft Matter Phys 68(2 Pt 1):021,915

[29] Coombes S, Hinch R, Timofeeva Y (2004) Receptors, sparks and waves in a fire-diffuse-fire framework for calcium release. Prog Biophys Mol Biol 85(2–3):197–216

[30] Cuthbertson KSR, Chay T (1991) Modelling receptor-controlled intracellular calcium oscillators. Cell Calcium 12:97–109

[31] Dash RK, Qi F, Beard DA (2009) A biophysically based mathematical model for the kinetics of mitochondrial calcium uniporter. Biophys J 96(4):1318–1332

[32] Dawson SP, Keizer J, Pearson JE (1999) Fire-diffuse-fire model of dynamics of intracellular calcium waves. Proc Natl Acad Sci U S A 96(11):6060–6063

[33] De Young G, Keizer J (1992) A single pool IP3-receptor based model for agonist stimulated Ca2+ oscillations. Proc Natl Acad Sci U S A 89:9895– 9899

[34] Desroches M, Guckenheimer J, Krauskopf B, Kuehn C (2012) Mixed-mode oscillations with multiple time scales. SIAM Review 54(2):211–288

[35] Domijan M, Murray R, Sneyd J (2006) Dynamical probing of the mechanisms underlying calcium oscillations. J Nonlin Sci 16(5):483–506

[36] Dufour JF, Arias I, Turner T (1997) Inositol 1,4,5-trisphosphate and calcium regulate the calcium channel function of the hepatic inositol 1,4,5-trisphosphate receptor. J Biol Chem 272:2675–2681

[37] Dupont G, Erneux C (1997) Simulations of the effects of inositol 1,4,5-trisphosphate 3-kinase and 5-phosphatase activities on Ca^{2+} oscillations. Cell Calcium 22(5):321–31

[38] Dupont G, Combettes L, Leybaert L (2007) Calcium dynamics: spatio-temporal organization from the subcellular to the organ level. Int Rev Cytol 261:193–245

[39] Dupont G, Combettes L, Bird GS, Putney JW (2011) Calcium oscillations. Cold Spring Harb Perspect Biol 3(3)

[40] Endo M (2009) Calcium-induced calcium release in skeletal muscle. Physiol Rev 89(4):1153–76

[41] Endo M, Tanaka M, Ogawa Y (1970) Calcium-induced release of calcium from the sarcoplasmic reticulum of skinned skeletal muscle fibres. Nature 228:34–36

[42] Falcke M (2003a) Buffers and oscillations in intracellular Ca^{2+} dynamics. Biophys J 84(1):28–41

[43] Falcke M (2003b) Deterministic and stochastic models of intracellular Ca^{2+} waves. New Journal of Physics 5:96

[44] Falcke M (2004) Reading the patterns in living cells – the physics of Ca^{2+} signaling. Advances in Physics 53(3):255–440

[45] Falcke M, Hudson JL, Camacho P, Lechleiter JD (1999) Impact of mitochondrial Ca^{2+} cycling on pattern formation and stability. Biophys J 77(1):37–44

[46] Fill M, Copello JA (2002) Ryanodine receptor calcium release channels. Physiol Rev 82(4):893–922

[47] Fill M, Zahradníková A, Villalba-Galea CA, Zahradník I, Escobar AL, Györke S (2000) Ryanodine receptor adaptation. J Gen Physiol 116(6):873–82

[48] Finch E, Goldin S (1994) Calcium and inositol 1,4,5-trisphosphate-induced Ca^{2+} release. Science 265:813–815

[49] Fletcher PA, Li YX (2009) An integrated model of electrical spiking, bursting, and calcium oscillations in GnRH neurons. Biophys J 96(11):4514–24

[50] Foskett JK, White C, Cheung KH, Mak DOD (2007) Inositol trisphosphate receptor Ca^{2+} release channels. Physiol Rev 87(2):593–658

[51] Friel DD (1995) $[Ca^{2+}]_i$ oscillations in sympathetic neurons: an experimental test of a theoretical model. Biophys J 68(5):1752–1766

[52] Gin E, Kirk V, Sneyd J (2006) A bifurcation analysis of calcium buffering. J Theor Biol 242(1):1–15

[53] Gin E, Crampin EJ, Brown DA, Shuttleworth TJ, Yule DI, Sneyd J (2007) A mathematical model of fluid secretion from a parotid acinar cell. J Theor Biol 248(1):64–80

[54] Goel P, Sneyd J, Friedman A (2006) Homogenization of the cell cytoplasm: the calcium bidomain equations. SIAM J Multiscale Modeling and Simulation 5:1045–1062

[55] Goldbeter A, Dupont G, Berridge M (1990) Minimal model for signal-induced Ca^{2+} oscillations and for their frequency encoding through protein phosphorylation. Proc Natl Acad Sci USA 87:1461–1465

[56] Greenstein JL, Winslow RL (2011) Integrative systems models of cardiac excitation-contraction coupling. Circ Res 108(1):70–84

[57] Greenstein JL, Hinch R, Winslow RL (2006) Mechanisms of excitation-contraction coupling in an integrative model of the cardiac ventricular myocyte. Biophys J 90(1):77–91

[58] Groff JR, Smith GD (2008) Calcium-dependent inactivation and the dynamics of calcium puffs and sparks. J Theor Biol 253(3):483–99

[59] Grubelnik V, Larsen AZ, Kummer U, Olsen LF, Marhl M (2001) Mitochondria regulate the amplitude of simple and complex calcium oscillations. Biophys Chem 94(1–2):59–74

[60] Haak LL, Song LS, Molinski TF, Pessah IN, Cheng H, Russell JT (2001) Sparks and puffs in oligodendrocyte progenitors: cross talk between ryanodine receptors and inositol trisphosphate receptors. Journal of Neuroscience 21(11):3860–3870

[61] Harvey E, Kirk V, Osinga HM, Sneyd J, Wechselberger M (2010) Understanding anomalous delays in a model of intracellular calcium dynamics. Chaos 20(4):045,104

[62] Harvey E, Kirk V, Wechselberger M, Sneyd J (2011) Multiple timescales, mixed mode oscillations and canards in models of intracellular calcium dynamics. J Nonlin Sci 21(5):639–683

[63] Higgins ER, Cannell MB, Sneyd J (2006) A buffering SERCA pump in models of calcium dynamics. Biophys J 91(1):151–63

[64] Hindmarsh JL, Rose RM (1984) A model of neuronal bursting using three coupled first order differential equations. Proc R Soc Lond B 221:87–102

[65] Hodgkin AL, Huxley AF (1952) A quantitative description of membrane current and its application to conduction and excitation in nerve. J Physiol 117(4):500–44

[66] Ilyin V, Parker I (1994) Role of cytosolic Ca^{2+} in inhibition of $InsP_3$-evoked Ca^{2+} release in Xenopus oocytes. J Physiol 477 (Pt 3):503–9

[67] Ionescu L, White C, Cheung KH, Shuai J, Parker I, Pearson JE, Foskett JK, Mak DO (2007) Mode switching is the major mechanism of ligand regulation of $InsP_3$ receptor calcium release channels. J Gen Physiol 130(6):631–45

[68] Izhikevich E (2000) Neural excitability, spiking and bursting. Int J Bif Chaos 10(6):1171–1266

[69] Jafri M (1995) A theoretical study of cytosolic calcium waves in Xenopus oocytes. J Theor Biol 172:209–216

[70] Janssen LJ, Kwan CY (2007) ROCs and SOCs: What's in a name? Cell Calcium 41(3):245–247

[71] Jasoni CL, Romano N, Constantin S, Lee K, Herbison AE (2010) Calcium dynamics in gonadotropin-releasing hormone neurons. Front Neuroendocrinol 31(3):259–69

[72] Jones C (1984) Stability of the traveling wave solutions of the FitzHugh-Nagumo system. Trans Amer Math Soc 286:431–469

[73] Keener J, Sneyd J (2008) Mathematical Physiology, 2nd edn. Springer-Verlag, New York

[74] Keizer J, Li YX, Stojilković S, Rinzel J (1995) InsP3-induced Ca^{2+} excitability of the endoplasmic reticulum. Mol Biol Cell 6(8):945–51

[75] Keizer J, Smith GD, Ponce-Dawson S, Pearson JE (1998) Saltatory propagation of Ca^{2+} waves by Ca^{2+} sparks. Biophys J 75(2):595–600

[76] Koch C, Segev I (eds) (1998) Methods in Neuronal Modeling; from Ions to Networks. MIT Press, Cambridge, MA

[77] Koivumaki JT, Takalo J, Korhonen T, Tavi P, Weckstrom M (2009) Modelling sarcoplasmic reticulum calcium ATPase and its regulation in cardiac myocytes. Philosophical Transactions of the Royal Society A: Mathematical, Physical and Engineering Sciences 367(1896):2181–2202

[78] Krupa M, Sandstede B, Szmolyan P (1997) Fast and slow waves in the FitzHugh-Nagumo equation. J Diff Eq 133(1):49–97

[79] Krupa M, Popovic N, Kopell N (2008) Mixed-mode oscillations in three time-scale systems: A prototypical example. SIAM J Appl Dyn Syst 7(2):361–420

[80] Krupa M, Vidal A, Desroches M, Clément F (2012) Mixed-mode oscillations in a multiple time scale phantom bursting system. SIAM J Appl Dyn Syst 11(4):1458–1498

[81] Kupferman R, Mitra P, Hohenberg P, Wang S (1997) Analytical calculation of intracellular calcium wave characteristics. Biophys J 72(6):2430–44

[82] LeBeau AP, Robson AB, McKinnon AE, Donald RA, Sneyd J (1997) Generation of action potentials in a mathematical model of corticotrophs. Biophys J 73(3):1263–75

[83] Lechleiter J, Girard S, Peralta E, Clapham D (1991) Spiral calcium wave propagation and annihilation in xenopus laevis oocytes. Science 252:123–126

[84] Lee K, Duan W, Sneyd J, Herbison AE (2010) Two slow calcium-activated afterhyperpolarization currents control burst firing dynamics in gonadotropin-releasing hormone neurons. J Neurosci 30(18):6214–24

[85] Leybaert L, Sanderson MJ (2012) Intercellular Ca^{2+} waves: mechanisms and function. Physiol Rev 92(3):1359–92

[86] Li YX, Rinzel J (1994) Equations for InsP$_3$ receptor-mediated Ca^{2+} oscillations derived from a detailed kinetic model: a Hodgkin-Huxley-like formalism. J Theor Biol 166:461–473

[87] Li YX, Rinzel J, Keizer J, Stojilković S (1994) Calcium oscillations in pituitary gonadotrophs: comparison of experiment and theory. Proc Natl Acad Sci USA 91:58–62

[88] Li YX, Rinzel J, Vergara L, Stojilković SS (1995) Spontaneous electrical and calcium oscillations in unstimulated pituitary gonadotrophs. Biophys J 69(3):785–95

[89] MacLennan DH, Rice WJ, Green NM (1997) The mechanism of Ca^{2+} transport by sarco(endo)plasmic reticulum Ca^{2+}-ATPases. J Biol Chem 272(46):28,815–8

[90] Maginu K (1985) Geometrical characteristics associated with stability and bifurcations of periodic travelling waves in reaction-diffusion equations. SIAM J Appl Math 45:750–774

[91] Magnus G, Keizer J (1997) Minimal model of beta-cell mitochondrial Ca^{2+} handling. Am J Physiol 273(2 Pt 1):C717–C733

[92] Magnus G, Keizer J (1998) Model of beta-cell mitochondrial calcium handling and electrical activity. i. cytoplasmic variables. Am J Physiol 274(4 Pt 1):C1158–C1173

[93] Mak DOD, Pearson JE, Loong KPC, Datta S, Fernández-Mongil M, Foskett JK (2007) Rapid ligand-regulated gating kinetics of single inositol 1,4,5-trisphosphate receptor Ca^{2+} release channels. EMBO Rep 8(11):1044–51

[94] Marchant JS, Parker I (2001) Role of elementary Ca^{2+} puffs in generating repetitive Ca^{2+} oscillations. EMBO J 20(1–2):65–76

[95] Marhl M, Haberichter T, Brumen M, Heinrich R (2000) Complex calcium oscillations and the role of mitochondria and cytosolic proteins. Biosystems 57:75–86

[96] Meinrenken CJ, Borst JG, Sakmann B (2003) Local routes revisited: the space and time dependence of the Ca^{2+} signal for phasic transmitter release at the rat calyx of Held. J Physiol 547(Pt 3):665–89

[97] Møller JV, Olesen C, Winther AML, Nissen P (2010) The sarcoplasmic Ca^{2+}-ATPase: design of a perfect chemi-osmotic pump. Quart Rev Biophys 43(04):501–566

[98] Neher E, Augustine GJ (1992) Calcium gradients and buffers in bovine chromaffin cells. J Physiol 450:273–301

[99] Neher E, Sakaba T (2008) Multiple roles of calcium ions in the regulation of neurotransmitter release. Neuron 59(6):861–72, DOI 10.1016/j.neuron.2008.08.019

[100] Neher EE (1995) The use of fura-2 for estimating Ca buffers and Ca fluxes. Neuropharmacology 34(11):1423–1442

[101] Palk L, Sneyd J, Shuttleworth TJ, Yule DI, Crampin EJ (2010) A dynamic model of saliva secretion. J Theor Biol 266(4):625–40

[102] Palk L, Sneyd J, Patterson K, Shuttleworth TJ, Yule DI, Maclaren O, Crampin EJ (2012) Modelling the effects of calcium waves and oscillations on saliva secretion. J Theor Biol 305:45–53

[103] Parekh AB, Putney JW (2005) Store-operated calcium channels. Physiol Rev 85(2):757–810

[104] Perez JF, Sanderson MJ (2005) The frequency of calcium oscillations induced by 5-HT, ACH, and KCl determine the contraction of smooth muscle cells of intrapulmonary bronchioles. J Gen Physiol 125(6):535–53

[105] Politi A, Gaspers LD, Thomas AP, Höfer T (2006) Models of IP$_3$ and Ca^{2+} oscillations: frequency encoding and identification of underlying feedbacks. Biophys J 90(9):3120–33

[106] Pradhan RK, Beard DA, Dash RK (2010) A biophysically based mathematical model for the kinetics of mitochondrial Na^+-Ca^{2+} antiporter. Biophys J 98(2):218–230

[107] Ressmeyer AR, Bai Y, Delmotte P, Uy KF, Thistlethwaite P, Fraire A, Sato O, Ikebe M, Sanderson MJ (2010) Human airway contraction and formoterol-induced relaxation is determined by Ca^{2+} oscillations and Ca^{2+} sensitivity. Am J Respir Cell Mol Biol 43(2):179–91

[108] Rinzel J (1985) Bursting oscillations in an excitable membrane model. In: Sleeman B, Jarvis R (eds) Ordinary and partial differential equations, Springer-Verlag, New York

[109] Romeo MM, Jones CKRT (2003) The stability of traveling calcium pulses in a pancreatic acinar cell. Physica D 177(1):242–258

[110] Roper P, Callaway J, Armstrong W (2004) Burst initiation and termination in phasic vasopressin cells of the rat supraoptic nucleus: a combined mathematical, electrical, and calcium fluorescence study. J Neurosci 24(20):4818–31, DOI 10.1523/JNEUROSCI.4203-03.2004

[111] Sala F, Hernàndez-Cruz A (1990) Calcium diffusion modeling in a spherical neuron: relevance of buffering properties. Biophys J 57:313–324

[112] Salido GM, Sage SO, Rosado JA (2009) TRPC channels and store-operated Ca^{2+} entry. BBA - Molecular Cell Research 1793(2):223–230

[113] Sanderson MJ, Bai Y, Perez-Zoghbi J (2010) Ca^{2+} oscillations regulate contraction of intrapulmonary smooth muscle cells. Adv Exp Med Biol 661:77–96

[114] Schuster S, Marhl M, Höfer T (2002) Modelling of simple and complex calcium oscillations. From single-cell responses to intercellular signalling. Eur J Biochem 269(5):1333–55

[115] Shannon TR, Wang F, Puglisi J, Weber C, Bers DM (2004) A mathematical treatment of integrated Ca dynamics within the ventricular myocyte. Biophys J 87(5):3351–71

[116] Shuai J, Pearson JE, Foskett JK, Mak DO, Parker I (2007) A kinetic model of single and clustered IP_3 receptors in the absence of Ca^{2+} feedback. Biophys J 93(4):1151–62

[117] Shuttleworth TJ (2012) STIM and Orai proteins and the non-capacitative ARC channels. Front Biosci 17:847–60

[118] Siekmann I, Wagner LE, Yule D, Crampin EJ, Sneyd J (2012) A kinetic model for type I and II IP_3R accounting for mode changes. Biophys J 103(4):658–68

[119] Simpson D, Kirk V, Sneyd J (2005) Complex oscillations and waves of calcium in pancreatic acinar cells. Physica D: Nonlinear Phenomena 200(3–4):303–324

[120] Skupin A, Falcke M (2009) From puffs to global Ca^{2+} signals: how molecular properties shape global signals. Chaos 19(3):037,111

[121] Smith G, Dai L, Miura R, Sherman A (2001) Asymptotic analysis of buffered calcium diffusion near a point source. SIAM J on Appl Math 61:1816–1838

[122] Smith GD (1996) Analytical steady-state solution to the rapid buffering approximation near an open Ca^{2+} channel. Biophys J 71(6):3064–72

[123] Smith GD, Wagner J, Keizer J (1996) Validity of the rapid buffering approximation near a point source of calcium ions. Biophys J 70(6):2527–39

[124] Smyth JT, Hwang SY, Tomita T, DeHaven WI, Mercer JC, Putney JW (2010) Activation and regulation of store-operated calcium entry. J Cell Mol Med 14(10):2337–2349

[125] Sneyd J, Dufour JF (2002) A dynamic model of the type-2 inositol trisphosphate receptor. Proc Natl Acad Sci USA 99(4):2398–403

[126] Sneyd J, Falcke M (2005) Models of the inositol trisphosphate receptor. Prog Biophys Mol Biol 89(3):207–45

[127] Sneyd J, Sherratt J (1997) On the propagation of calcium waves in an inhomogeneous medium. SIAM J Appl Math 57:73–94

[128] Sneyd J, Dale P, Duffy A (1998) Traveling waves in buffered systems: applications to calcium waves. SIAM J Appl Math 58:1178–1192

[129] Sneyd J, Tsaneva-Atanasova K, Yule DI, Thompson JL, Shuttleworth TJ (2004) Control of calcium oscillations by membrane fluxes. Proc Natl Acad Sci USA 101(5):1392–6, DOI 10.1073/pnas.0303472101

[130] Sneyd J, Tsaneva-Atanasova K, Reznikov V, Bai Y, Sanderson MJ, Yule DI (2006) A method for determining the dependence of calcium oscillations on inositol trisphosphate oscillations. Proc Natl Acad Sci USA 103(6):1675–80, DOI 10.1073/pnas.0506135103

[131] Soboloff J, Rothberg BS, Madesh M, Gill DL (2012) STIM proteins: dynamic calcium signal transducers. Nat Rev Mol Cell Biol 13(9):549–565

[132] Soeller C, Cannell MB (2004) Analysing cardiac excitation-contraction coupling with mathematical models of local control. Prog Biophys Mol Biol 85(2–3):141–62

[133] Straub SV, Giovannucci DR, Yule DI (2000) Calcium wave propagation in pancreatic acinar cells: functional interaction of inositol 1,4,5-trisphosphate receptors, ryanodine receptors, and mitochondria. J Gen Physiol 116(4):547–560

[134] Tanimura A, Morita T, Nezu A, Tojyo Y (2009) Monitoring of IP_3 dynamics during Ca^{2+} oscillations in HSY human parotid cell line with FRET-based IP_3 biosensors. J Med Invest 56 Suppl:357–61

[135] Thomas D, Lipp P, Tovey SC, Berridge MJ, Li W, Tsien RY, Bootman MD (2000) Microscopic properties of elementary Ca^{2+} release sites in non-excitable cells. Curr Biol 10(1):8–15

[136] Thul R, Bellamy TC, Roderick HL, Bootman MD, Coombes S (2008) Calcium oscillations. Adv Exp Med Biol 641:1–27

[137] Thurley K, Skupin A, Thul R, Falcke M (2012) Fundamental properties of Ca^{2+} signals. Biochim Biophys Acta 1820(8):1185–94

[138] Toyoshima C (2008) Structural aspects of ion pumping by Ca^{2+}-ATPase of sarcoplasmic reticulum. Arch biochem biophys 476(1):3–11

[139] Tsai J, Sneyd J (2005) Existence and stability of traveling waves in buffered systems. SIAM J Appl Math 66(1):237–265

[140] Tsai JC, Sneyd J (2007) Are buffers boring? Uniqueness and asymptotical stability of traveling wave fronts in the buffered bistable system. J Math Biol 54(4):513–53, DOI 10.1007/s00285-006-0057-3

[141] Tsai JC, Sneyd J (2011) Traveling waves in the buffered FitzHugh-Nagumo model. SIAM J Appl Math 71(5):1606–1636

[142] Tsai JC, Zhang W, Kirk V, Sneyd J (2012) Traveling waves in a simplified model of calcium dynamics. SIAM J Appl Dyn Syst 11(4):1149–1199, DOI 10.1137/120867949

[143] Tsaneva-Atanasova K, Osinga HM, Riess T, Sherman A (2010a) Full system bifurcation analysis of endocrine bursting models. J Theor Biol 264(4):1133–46

[144] Tsaneva-Atanasova K, Osinga HM, Tabak J, Pedersen MG (2010b) Modeling mechanisms of cell secretion. Acta Biotheoretica 58(4):315–327

[145] Tuan HT, Williams GS, Chikando AC, Sobie EA, Lederer WJ, Jafri MS (2011) Stochastic simulation of cardiac ventricular myocyte calcium dynamics and waves. Conf Proc IEEE Eng Med Biol Soc 2011:4677–80

[146] Ventura AC, Sneyd J (2006) Calcium oscillations and waves generated by multiple release mechanisms in pancreatic acinar cells. Bull Math Biol 68(8):2205–31, DOI 10.1007/s11538-006-9101-0

[147] Wagner J, Keizer J (1994) Effects of rapid buffers on Ca^{2+} diffusion and Ca^{2+} oscillations. Biophys J 67:447–456

[148] Wagner LE, Yule DI (2012) Differential regulation of the $InsP_3$ receptor type-1 and -2 single channel properties by $InsP_3$, Ca^{2+} and ATP. J Physiol 590(14):3245–3259

[149] Wang IY, Bai Y, Sanderson MJ, Sneyd J (2010) A mathematical analysis of agonist- and KCl-induced Ca^{2+} oscillations in mouse airway smooth muscle cells. Biophys J 98(7): 1170–1181

[150] Wechselberger M (2012) A propos de canards (apropos canards). Trans Amer Math Soc 364:3289–3309

[151] Williams GS, Chikando AC, Tuan HT, Sobie EA, Lederer WJ, Jafri MS (2011) Dynamics of calcium sparks and calcium leak in the heart. Biophys J 101(6):1287–96

[152] Williams GSB, Molinelli EJ, Smith GD (2008) Modeling local and global intracellular calcium responses mediated by diffusely distributed inositol 1,4,5-trisphosphate receptors. J Theor Biol 253(1):170–88

[153] Winslow RL, Greenstein JL (2013) Extinguishing the sparks. Biophys J 104(10):2115–7

[154] Yao Y, Choi J, Parker I (1995) Quantal puffs of intracellular Ca^{2+} evoked by inositol trisphosphate in Xenopus oocytes. J Physiol 482 (Pt 3):533–53

[155] Yule DI (2010) Pancreatic acinar cells: molecular insight from studies of signal-transduction using transgenic animals. Int J Biochem Cell Biol 42(11):1757–61

[156] Zhang M, Goforth P, Bertram R, Sherman A, Satin L (2003) The Ca^{2+} dynamics of isolated mouse beta-cells and islets: implications for mathematical models. Biophys J 84(5): 2852–2870

[157] Zhang W, Kirk V, Sneyd J, Wechselberger M (2011) Changes in the criticality of Hopf bifurcations due to certain model reduction techniques in systems with multiple timescales. J Math Neurosci 1(1):9

[158] Zhang W, Krauskopf B, Kirk V (2012) How to find a codimension-one heteroclinic cycle between two periodic orbits. Discrete Cont Dyn S 32(8):2825–2851